Sound, Noise, and Vibration Control

Sound, Noise, and Vibration Control

Second Edition

LYLE F. YERGES

Consulting Engineer

ROBERT E. KRIEGER PUBLISHING COMPANY
MALABAR, FLORIDA

Second Edition   1978
Reprint Edition   1983, 1985 , 1987

Printed and Published by
ROBERT E. KRIEGER PUBLISHING COMPANY, INC.
KRIEGER DRIVE
MALABAR, FLORIDA 32950

Library of Congress Cataloging in Publication Data

Yerges, Lyle F.
    Sound, noise, and vibration control.

    Reprint. Originally published: 2nd ed. New York :
Van Nostrand Reinhold, 1978. (Van Nostrand Reinhold
environmental engineering series)
    Bibliography: p.
    Includes index.
    1. Soundproofing. 2. Buildings—Vibration. 3. Acous-
tical engineering. I. Title.
[TH1725.Y47 1983]   693.8'34   83-11303
ISBN 0-89874-654-X

10 9 8 7 6 5 4 3

Introduction to Second Edition

Almost ten years ago, the first edition of *Sound, Noise, and Vibration Control* was conceived as a part of the Environmental Engineering Series. At that time, acoustics was just emerging from a minor place in architecture. Its esoteric language and the mystique which surrounded its practice tended to confuse the public and to make even engineers wary of it. The well-publicized remark of the disappointed architect of Philharmonic Hall at Lincoln Center—"Acoustics is a very inexact science"—summed it up pretty well for most people in 1967.

For decades, architects and possibly musicians were the only professionals whose work forced them to have at least a speaking knowledge of acoustics. For the past decade, however, and particularly since OSHA became a part of the lives of most working people, noise and its control have become vitally important to plant engineers, owners, machine designers, and even company treasurers.

Now noise ranks with the potential for skin cancer as one of the alleged hazards of operation of supersonic airplanes. Only ten years ago, the fascination with flying faster than the speed of sound precluded serious consideration of the acoustical penalties involved.

During the past decade, however, our growing concern with the environment and with occupational health and safety has sharply changed the emphasis of sound control as a discipline. With the Walsh–Healey Act of 1969 and the subsequent OSHA Act of 1971, decibels and acoustical criteria became a part of our everyday language: Municipal ordinances, building codes, environmental impact statements, and product performance standards began to include acoustical requirements and noise limits. Congressional committees and new government agencies wrestled with the complexities of sound and its effects on man. Engineers, plant managers, safety directors, and even lawyers and corporate officials flocked to seminars and short courses all over the nation.

New criteria, descriptors, and standards proliferated, until today a bewildering vocabulary has emerged, leaving all but the most sophisticated practitioners more confused than before.

Because acoustics is, or should be, an essential part of the language of practicing professionals, we have revised *Sound, Noise, and Vibration Control*, shifting the emphasis from its architectural and building design applications to its broader environmental scope. Sound control is important not just in concert halls and hi-fi set design, but in nearly all human environments, including the backyard patio and the factory.

An author must always ask himself, "Why bother to write this book at all?" This is particularly true in the field of acoustics, where a flood of books since 1970 includes scholarly texts on physical acoustics and even publications alleging to divulge for the first time the "secrets of noise control."

Every attempt has been made in this book to avoid scholarly language and to dispel the notion that there are any particular secrets or mystique involved in acoustics. Really, the principles are simple and clear. If it were not for the esoteric language of acoustics, nearly any competent professional, particularly any competent engineer, would be able to handle the subject in an almost routine manner. However, acousticians appear determined to make the subject inaccessible, even to themselves. Almost yearly they spawn a new collection of criteria, symbols, terms, procedures, and standards. It is ironic that after almost a quarter century of research and discussion which has developed a flood of new terms and criteria intended to make accurate, meaningful, objective determination of the subjective effects of sound, we still find the A-weighted network of an ordinary sound level meter one of the best and simplest means of identifying human reactions to sound.

This is unabashedly a cookbook. It is written on the premise that few people find the subject of acoustics of compelling interest; but, like it or not, many people must do something about the acoustics of the spaces in which they live and work.

The highest accolade repeatedly accorded to the first edition was that it was "practical." That is what we hoped to accomplish then and what we strive for now—to reduce to *practice* what we have learned about the science of acoustics.

Science deals with disciplined observation and developing theories which are attempts to explain what is repeatedly observed. It sets up mathematical models and analytical procedures, all limited by the accuracy and precision of measurements and observations and by our ability to handle the mathematics involved.

Insofar as possible, this book uses an intuitive, pragmatic approach in preference to the excessively analytical. We have tried to avoid today's tendency to set up computer programs and then to beat the problem to death, often without insight or understanding. Rules of thumb, reasonable and time-tested values, tables of empirical data, and simplified calculations are used whenever possible.

Simplified measurement procedures, standard good practice, and reasonable ranges of performance have been used, even at the possible sacrifice of precision. In a science where one dB (very difficult to measure) represents about 26% change, and where three dB (the change necessary to be even noticeable to most listeners in normal circumstances) represents a two-to-one ratio in energy, it seems difficult to justify hair-splitting or expensive and time-consuming calculations, research, and investigation.

As the title of the book implies, we are dealing with the *control* of sound, noise, and vibration. Sound is a very essential part of our lives. Eliminating it would normally be a tragedy, making useless one of man's finest senses and isolating him from some of the most important stimuli in the universe. Noise, on the other hand, is simply unwanted sound—unwanted for any reason. Eliminating that part of sound from our lives is important.

Vibration, as that word is normally used, represents the counterpart of audible sound and is normally felt or experienced independently of the ears.

Insofar as possible, the organization of this book is such that essential aspects of sound, noise, and vibration are associated with feasible means of controlling them in order to make a safer, more hospitable environment for humans. No plant engineer needs be reminded that noise control now has the compulsion of the law to give it importance. Nearly every hearing on proposed legislation for noise control is well attended and widely publicized. No longer does anyone dismiss noise as an unimportant but necessary part of modern life. Quite the contrary, in some instances almost too much emphasis is put upon the problem, an almost inevitable result of emotional and ideological response rather than intelligent, considered appraisal of the facts.

The contents of this book represent the distillation of a lifetime of experience in the field of acoustics and the thousands of pages of publications devoted to the subject. As a result, it probably suffers from a certain arbitrariness of tone and a somewhat simplistic approach to what are really complex problems. It cannot hope to answer all questions or to solve all aspects of any problem. Just as a cookbook is not an expert treatise on nutrition nor a dissertation on the refinements of gourmet cuisine, this book is not intended to produce experts. Rather, it is aimed toward leading the uninitiated into developing necessary skills in an important activity and to guide the novice in becoming better and more effective in his work.

About Metric or SI Units

We debated at length about whether to go completely to SI units, stay with English units, or use a mixture of both. As might be expected, we decided on a compromise, one which we believe will be most "practical" for most U.S. readers.

Since most reference books, data collections, and specifications available to U.S. readers are in English units, we chose English units for most of the tables and graphs in the book. It is unlikely that those reference sources will be converted to SI units for at least one or more decades.

Tables of conversion factors are included to simplify the task of translating the English units to SI.

In the body of the text, SI or metric units are either included in parentheses or parallel columns of tables.

In any future edition of the book, it is likely that SI units will be used exclusively.

Preface

This book deals with *human* factors—the factors which affect the environment of man. Sometimes it seems as if this aspect of science and technology is almost forgotten, as if these terms mean only increasing power, gigantic scale, and enormous output. Often it appears that the purpose for all this activity—humanity—has been overlooked in the frantic pursuit of other, less important goals.

A society which will seriously consider airplanes that can blanket a fifty-mile-wide strip of the country below them with a boom more devastating than the crash of thunder, a society which will pollute its streams and lakes, foul the air it breathes, and brutalize the environment in which it lives is not necessarily an advanced society.

For the most part, humanity and the quality of human life provide our *raison d'être* as professionals. Usually people are not so much interested in the things we design or build as in the effect of such things upon the environment in which man lives and the way in which he lives in that environment. The principal activity of most professionals is to control and manipulate man's environment intelligently. This suggests identifying, quantifying, and qualifying the important environmental factors.

Although he may be embarrassed to admit it, the architect's or engineer's world today is a world of signs and symbols. A vast majority of practicing professionals rarely, if ever, handle, touch, or manipulate the actual objects which they claim to build or design. In fact, it is possible for an engineer never to see any of the "things" with which he ostensibly "works." The vocabulary of his profession consists of symbols and labels—a form of shorthand which he uses to communicate his ideas or concepts to others who must use them. He speaks of "measuring, evaluating, specifying, testing," He uses the symbolic language of mathematics, increasingly esoteric instruments and test procedures, and, more than ever, the mysterious thing called a computer to produce reams of data which have meaning only to the initiated. Thus, the term "professional" has come to imply a strange other world, somewhat remote from the everyday world of reality.

Unfortunately, this alienation from reality can become a serious handicap to the effective functioning of a professional in his own sphere of professional activity. The title of this book, for example—*Sound, Noise, and Vibration Control*—implies a host of concepts rarely understood, even by many of the practitioners in the field of sound and vibration. We have tried to translate these terms into a language more meaningful to everyone and to interpret what is meant by the symbols used.

We are immersed in a universe of energy, energy in various forms, manifesting itself in innumerable ways. Our environment is determined by the manifestations of this energy. Heat, light, and sound are three of the most common and important forms of energy in our environment. Nearly everyone assumes that he understands what these three words mean. After all, everyone knows when he sees something, when he feels hot or cold, or when he hears something. However, although we may be familiar with the physical or subjective effects of these forms of energy, we may not have any idea of the nature of the energy itself.

Sound is so much a part of our environment, such a vital part of our lives, that one might assume that it would be taught, learned, analyzed, and understood from early school years on. Architects and engineers, particularly, might be expected to be especially familiar with the nature of sound and the techniques of sound control. Yet there is a lamentable ignorance of the subject, even among professionals.

Unfortunately, even in a day of highly advanced technology, rarely is the acoustical performance of a machine or a building the result of careful original planning and design. Much more often, sound and vibration control (or, more popularly, "noise control") are afterthoughts—something "stuck on" or added as an afterthought or in a desperate attempt to mitigate a serious and unforeseen problem.

From a career which includes almost forty years' experience in consulting, teaching, product development, and design and supervision of construction I have gleaned some clues to the reasons for this situation. An examination of physics textbooks used at all levels of formal education shows that little attention is given to the subject of sound; and when it is, the information is often trivial, unconnected with the real world of sound and acoustics, or even incorrect. On the other hand, there is a wealth of excellent books, journals, and other specialized publications which deal at length with physical acoustics and other aspects of sound. However, the language of these publications is often esoteric, full of complex mathematics, and not particularly informative, even to professionals in other fields.

It is not my intention to contribute to the alleged "information explosion," if, indeed, such a thing is occurring (it seems to me that a "publication explosion" is apparent, but one can seriously question how much information is actually involved in the phenomenon). This book has a specific and limited purpose—to provide, in the idiom of the practicing architect and engineer, enough of the fundamentals of sound and vibration and their control to permit the professional to feel comfortable about the subjects. Then, perhaps, man's acoustical environment eventually may contribute to the quality of his life rather than adding to the burden of modern living.

The purpose of this book is consistent with the title of the series—"Environmental Engineering." It proceeds from the *subjective*—the human factors—to the *objective*—measurements, definitions, and solutions to the human problems. Designs, constructions, systems, and materials are only means of meeting the subjective requirements. No one ever buys a drill—he wants a hole, and the drill is simply a means of providing the hole. We want an environment in which our subjective needs or desires can be fulfilled best.

To a large degree, many of the environmental parameters which affect the quality of our lives have been defined, at least to the point where we are able to deal with these subjective matters in an objective way.

As in all human endeavors, there is a hierarchy of importance and a priority of needs. Reluctantly, we admit that almost all solutions to problems are compromises with available resources, the state of the art, and the established prejudices of the individuals involved. This book attempts to recognize these practical realities and to make alternatives available to the designer. Furthermore, we hope that it will help him to forecast the consequences of all decisions before the results of these decisions are materialized in a machine, a structure, or a building. After all, a plan is simply a group of decisions made in advance; and a program is a predetermined sequence of carrying out the decisions. If the results of the decisions can be known throughout the planning and programming process, there need be no major surprises when the project is completed.

True, acoustics is still as much in the realm of "art" as of "science." But this need not be unfortunate; rather, it should give a competent and creative practitioner the opportunity to exercise options and to employ his creativity in an intelligent manner.

This book is intended to be a working guide for the professional. Unfortunately, too often the professional turns to sources and references only when he is in trouble or after his design has progressed to where changes or modifications are difficult, expensive, or impossible. It is our hope that this entire volume will create in the professional an attitude, an approach, and a concept. We hope to transform "form follows function" from a cliche into a meaningful guide for design. All too often the professional is "a solution going about looking for a problem." Preconceived notions, prejudices, patterns of thought and behavior have so conditioned him that he finds himself superimposing little snatches of solutions on to an already highly alienated design. This can be disastrous.

Certainly a knowledge of acoustics and skill in sound control should not hamper the designer. We have never found that ignorance solves any problems, nor that unawareness enhances creativity or productivity. It is still the professional's obligation to make decisions among alternatives, even when a measure of doubt exists. It is our intention only to minimize that doubt and to reduce the probability of disaster inherent in design processes which proceed from ignorance to ignorance.

HOW THIS BOOK IS ORGANIZED

This book is divided into three principal parts:

1. The basic, essential theory of sound and vibration necessary for a real understanding of the effect of this form of energy on people and the environment in which they live and work.
2. The broad, general principles of sound and vibration control, including the types of materials, systems, and constructions used for this purpose.
3. Important data, organized into tables, detailed drawings and sketches, and checklists for easy reference.

In general, the information includes well-established and widely used practices and procedures, universally available materials and constructions, and the methods distilled from long experience in the field. We are tempted to say that the ideas are just "good, common sense," but we know that "common sense" and "good sense" are not necessarily synonymous.

We know that some readers will rarely use anything but the third section of the book; they will want only a quick-reference handbook for solutions to immediate problems.

Some readers may want to be familiar with the principles which underlie sound control techniques and procedures, and some readers may even want to understand something of the nature of the ubiquitous energy surround which is such a vital part of our world.

We hope the book will be useful to all three groups—and even to others not so classified.

We have not derived equations nor dealt extensively with the mathematics associated with the phenomena described. Simple formulas, such as those associated with Noise Reduction and Transmission Loss, are stated and, where possible, their source quoted. ASTM test methods and standards are cited whenever they are essential.

In Section III, each table refers back to the appropriate pages of Section I or II, so that readers interested in the background for each table or collection of data can find it in the text of previous sections.

All material in Section II (Sound Control) and in the tables and charts of Section III is arranged in the order in which the practicing professional would approach any acoustical problem, beginning with a determination of the functions and criteria for the building spaces, the building site, layout and orientation of spaces, etc., through final tests and inspection (just as an acoustical consultant would approach a project if he were involved from the beginning).

It is demonstrated, for example, how certain acoustical criteria are fulfilled almost automatically by proper functional design and shaping of a space. It will become apparent that some of the acoustical criteria often overemphasized in older books on architectural acoustics naturally evolve from good design; they are not part of a mystique nor are they the result of the superimposed skill of an esoteric branch of the scientific fraternity.

No attempt has been made to discuss every conceivable room in any building or even the principal kinds of rooms in the more common buildings. Rather we have included those important types of spaces which illustrate in their design the major principles of sound and vibration control.

This book does not provide a step-by-step procedure for designing a symphony hall, for example. Such a building deserves the careful and loving attention not only of the architect but of an experienced and skillful acoustical consultant; its design cannot be covered in even a large book. While it is possible to find in books somewhere a wealth of advice and suggestions for designing almost any type of space, such activity does not appear to be the most fruitful way for a designer to employ his limited time. Some projects simply require the services of the best experts available. Few architects or engineers will ever design a concert hall, a major radio or television studio, or even a large theater or sports arena. When and if they do, it may be the part of wisdom to plan on the assistance of a consultant from the outset. Still, acoustics is a vital part of even the simplest, most commonplace space, including the factory. The architect and the engineer should not only be aware of this, but they should be reasonably skilled in handling the normal, ordinary problems associated with good acoustical design and sound control.

To simplify use of the book, the usual footnotes and references have been omitted. We hope the reader will assume that we have read the references and can cite the "authorities" if necessary. A general list of journals, books, and other sources is included in the bibliography and reference list. It is virtually impossible to include all of the literally thousands of sources from which the information is drawn. Inclusion or omission of any source is not necessarily an assessment of its worth. Rather, we have included the sources which we have found particularly useful to us in our own work. Each of the references listed usually contains its own lengthy list of sources and additional bibliography.

In general, we have not included collections of data which may either become quickly obsolete or which are provided regularly and currently by trade associations, professional organizations, or technical societies. The titles and sources of the more significant collections of such data are listed.

The collections of data included are brief but reasonably inclusive; for most applications, they are adequate. For highly specific applications, it is always wise to obtain the most current, specific data available, either directly from a manufacturer or from a current listing published by a reputable source.

ACKNOWLEDGMENTS

It is difficult to acknowledge properly my debt to the innumerable contributors —intentional or unintentional—to this book. In a career of almost forty years, one assembles, assimilates, and uses data from so many sources that it is not possible to be sure what came from where or whom. On the premise that most originality is undetected plagiarism, I wish to thank my colleagues and associates from whom I learned and accumulated the information on which this book is based.

To my former associates during my many years as an officer or committee member of the Acoustical Materials Association, I wish to extend sincere thanks for their generous contributions of information and advice. Manufacturers of acoustical materials and devices have been extremely cooperative and helpful. If we have forgotten or overlooked any of them, we extend sincere apologies.

I am particularly grateful to Mrs. Barbara Wilson who typed the entire manuscript for the first edition, prepared and assembled the tables and collections of data, edited and corrected the text, and, in general, did the important work associated with "writing" and to Mrs. Kathryn Wynes who performed the same tasks for the second edition.

And to my family, who tolerated the lost nights and weekends for many months, and who gave up a dining room for that entire period (the table was piled high with references, manuscripts, illustrations, etc.), my thanks for their patience and understanding.

LYLE F. YERGES

Contents

Introduction to Second Edition   v
Preface   ix
How This Book Is Organized   xii
Acknowledgments   xiv

SECTION I

1   The Nature of Sound   1
2   Hearing   9

SECTION II

3   Controlling Sound, Noise, and Vibration   17

SECTION III

Design Procedure   84
Background Noise Criteria   88
Other Criteria   90
Site Planning   91
Layout and Orientation of Spaces   93
Typical Noise Sources   94
Choosing the Exterior Construction   95
Choosing the Interior Construction   96
    Sound Transmission Loss–Partitions and Walls   104
    Sound Absorption   131
    Shape and Configuration   137
    Sound Reinforcement   148
Mechanical Equipment Noise and Vibration Control   153
    Noise and Vibration Control   157
Heating, Piping, Air Conditioning, and Electrical Systems   170
    General   170
    Fan and Duct Systems   170
    Piping Systems   178
    Electrical Equipment   181

Other Mechanical Equipment   182
Other Sound Control Devices   183
Sound Fields and Sound in Enclosures   184
  General   184
  Sound Fields   184
  Sound in Enclosures   185
  Environmental Noise Control   187
  Industrial Noise Control   190
Legal and Medical Problems   194
  Hearing Conservation Criterion   194
  Vibration Criteria   195
  Effect of Noise on Productivity   197
  Zoning and Noise Ordinances   198
  Personal Protection Measures   198
Tests and Measurements   201
  Laboratory Tests   201
  Field Tests   201
  Field Measurements   204
Trouble Shooting   206
  Echo, Flutter, Reverberation, and Focusing Effects   209
  Transmission   209
  Vibration   210
  Source-Path-Receiver   211
Industrial Noise Control—Case Histories   213
  General   213
  Cost/Benefit Analysis   213
  Mock-ups   214
  Case Histories   214
Glossary   227
Bibliography and References   233
The International System of Units (SI)   237

INDEX   240

List of Illustrations

Figure 1	Motion of molecules in elastic medium as a sound wave progresses through medium.	2
Figure 2	Motion of a molecule during a single cycle.	3
Figure 3	An elastic medium subjected to sound.	3
Figure 4	Equal loudness contours.	11
Figure 5	"Voice-print" of an office calculating machine.	13
Figure 6	Human response to vibration.	15
Figure 7	Frequency response for sound level meter networks.	20
Figure 8	Relationship between sones and phons.	21
Figure 9	Equal "noisiness" contours.	22
Figure 10	Noise criteria curves.	23, 87
Figure 11	A wall as a "sound barrier."	32, 95
Figure 12	How sound moves a "sound barrier."	33
Figure 13	Comparison between Transmission Loss of "limp mass" and actual panel.	36
Figure 14	Comparison between Transmission Loss of "limp mass" and same mass divided into double wall.	37
Figure 15	Determination of Sound Transmission Class.	38, 97
Figure 16	Determination of Impact Noise Rating of floors.	41, 103
Figure 17	The effect of leaks.	42, 121
Figure 18	Transmission by flanking.	43
Figure 19	Flanking through acoustical ceilings.	44
Figure 20	Photomicrograph of open-cell foamed urethane.	47
Figure 21	Diffraction through a perforated facing.	49
Figure 22	Effect of mounting method.	51
Figure 23	Sound absorption of panel materials.	54
Figure 24	Concrete block resonator.	55
Figure 25	Absorption of concrete block resonator.	56
Figure 26	Decay of a damped oscillation.	61
Figure 27	A simple vibrating system with damping.	61
Figure 28	Transmissibility of a vibrating system.	63
Figure 29	Dynamic response vs. static response of a typical "high-damping" elastomer.	64
Figure 30	Large multi-cell horn.	72
Figure 3-1	Sound Pressure Level in dBA of typical sounds.	85
Figure 3-2	Significant frequency ranges.	86
Figure 3-3	Site planning.	91
Figure 3-4	Orientation on site.	91
Figure 3-5	Space planning for acoustical privacy.	93

Figure 3-6   The effect of caulking.   122
Figure 3-7   Caulking details.   125
Figure 3-8   Typical leaks and flanking sources.   126
Figure 3-9   Typical flanking paths.   127
Figure 3-10   Flanking via the ceiling.   128
Figure 3-11   Avoiding ceiling flanking.   128
Figure 3-12   Optimum reverberation time.   135
Figure 3-13   Reflection of sound.   139
Figure 3-14   Echo.   140
Figure 3-15   Flutter.   140
Figure 3-16   Shaping Special Spaces.   143
Figure 3-17   Shaping Music Rehearsal Spaces.   144
Figure 3-18   Shaping Auditorium-type Spaces.   145
Figure 3-19   "Throw-ratio" for speakers.   149
Figure 3-20   Speaker location and coverage.   150
Figure 3-21   Minimizing structural transmission of noise and
                  vibration.   159
Figure 3-21b   Typical vibration criteria for mechanical equipment.   159
Figure 3-22a   Effect of sound control measures.   162
Figure 3-22b   Effect of sound control measures.   163
Figure 3-23   Required static deflection for resilient mounts.   165
Figure 3-24   Vibration isolation system for mechanical equipment.   171
Figure 3-25   Effect of damper closure.   172
Figure 3-26   Effect of changes in operating conditions on perfor-
                  mance of typical diffusers.   173
Figure 3-27   Typical flanking paths.   176
Figure 3-28   Resilient connection system for small pipelines.   178
Figure 3-29   Flexible pipe connector with isolated restraining bolts.   179
Figure 3-30   Hearing conservation criterion.   195
Figure 3-31   Vibration exposure criteria.   197
Figure 3-32   Simplified Field Sound Transmission test.   202
Figure 3-CH-1   Conveyor lines.   215
Figure 3-CH-2   Centrifugal pump.   216
Figure 3-CH-3   Hydraulic power unit.   217
Figure 3-CH-4   High-speed stamping press.   218
Figure 3-CH-5   High-speed stamping presses.   219
Figure 3-CH-6   Pulverizer.   220
Figure 3-CH-7   Run-up room for chain saws.   221
Figure 3-CH-8   Foundry chipping booths.   223
Figure 3-CH-9   Welding assembly line.   224
Figure 3-CH-10   Automatic screw machine plant.   225
Figure 3-CH-10b   Noise levels; automatic screw machine plant.   226

List of Tables

Table 1	Velocity of Sound in Various Media	5, 84
Table 2	Subjective Effect of Changes in Sound Characteristics	7, 86
Table 3	Comparison of Intensity, Sound Pressure Level, and Common Sounds	8
Table 3a	Decibel Notation in Acoustics	8
Table 3b	Decibel Addition	9
Table 4	Subjective Responses to Characteristics of Sound	10
Table 5	Significant Frequency Ranges	12
Table 6	Sound Transmission Loss for Common Building Materials	39
Table 7	Common Porous Absorbents	52, 132
Table 8	Acoustical Impedance of Various Materials	59
Table 3-1	Spaces	88
Table 3-2	Neighborhoods	89
Table 3-2b	Other Criteria	90
Table 3-3	Effect of Distance	92
Table 3-4	Typical Noise Sources	94
Table 3-5a	Wall, Partition, or Panel Between	98
Table 3-5b	Wall, Partition, or Panel Between	100
Table 3-6	Impact Noise Criteria for Floors	104
Table 3-7	STC-ratings of Partitions and Walls	105
Table 3-8	STC-ratings and INR-ratings of Floors	114
Table 3-9	Amount, Type, and Location of Absorption	133
Table 3-10	Mechanical Equipment Noise	154
Table 3-10b	Effect, in dB, of Operating Parameters on Machine Noise	158
Table 3-11	Mounting System Requirements	169
Table 3-11b	Room Effect—in dB	186
Table 3-11c	Industrial Noise Exposure Criteria	196
Table 3-11d	Typical Environmental Noise Criteria	199
Table 3-11e	Difference Between Linear and A-weighted Level	205
Table 3-12	Typical Acoustical Problems	206
Table 3-13	Available Options in Noise Control	212
Table 3-14	Industrial Noise Control Cost/Benefit Analysis	214

Section I

1. THE NATURE OF SOUND

Sound is a vibration in an elastic medium.

Sound is a relatively simple form of energy, causing variations in pressure and alternations in direction of molecular movement within media. Sound, like all objective things, exists by definition, and would exist even without any living thing to receive it or to be affected by it.

Acoustic energy is the total energy of a given part of a transmitting medium, minus the energy which would exist in the same part of the medium with no sound present. The transmission and dissipation of acoustic energy are of interest to nearly everyone involved in engineering and architectural activities.

Vibration refers to the oscillating motion of media.

The term "sound" is usually applied to the form of energy which produces hearing in humans, while "vibration" usually refers to the nonaudible acoustic phenomena which are recognized by the tactile experience of touch or feeling. However, there is no essential difference between the sonic and vibratory forms of sound energy..

Source–Path–Receiver

Sound originates with a *source*—some energy input of some sort; travels via a *path*—an elastic medium of some type; and reaches a *receiver*—usually the human body is the receiver of interest to us, although animals and equipment can be "receivers," too.

Understanding the nature of sound propagation is imperative if the principles of sound control are to be correctly understood. So many misconceptions about the subject persist that an examination of the basic mechanism of sound generation and transmission is fundamental.

Sound can be generated in almost anything with which we work and in almost any normal environment in which we live. Obviously, in the almost empty

1

reaches of space there can be no sound, and vocal conversation between persons would be impossible. The space capsules of our astronauts are actually little islands of our world in which the environment of our world is reproduced sufficiently to permit maintenance of human life and activities, including speaking and hearing.

Sound Propagation in Elastic Media

We must understand the nature of "elastic matter" before we can understand what takes place as the sound motion occurs. Usually elastic media are those in which stress is proportional to strain. Molecules in any substance are constantly moving at a furious rate, depending upon temperature and pressure in the medium. They are striking one another, rebounding and striking other particles. Sound motion is superimposed on this already existing motion. For the purposes of this discussion, however, we will take an imaginary instant in time where all particles are equally spaced, and we will call their position at this instant "rest."

If we were to isolate a few molecules, approximately this sequence would be observed:

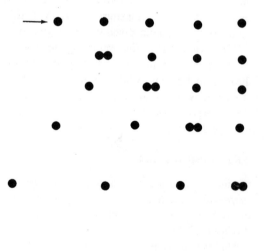

a. Molecules at "rest"; energy is applied to first.
b. First molecule strikes second.
c. Second strikes third; first starts slowly back toward its original position.
d. Third strikes fourth; second starts slowly back toward its original position; first back at original position.
e. Fourth strikes fifth; third starts slowly back toward its original position; second back at original position; first now stopped and at maximum position beyond its original position.

Figure 1 Motion of molecules in elastic medium as a sound wave progresses through medium.

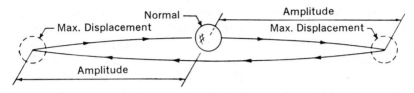

Figure 2 Motion of a molecule during a single cycle.

Figure 2 shows the motion of the first molecule. It was displaced first to one side of its normal "rest" position, then back through normal to the opposite position, and then back to normal. If this motion is regular and repetitive, it is called a vibration.

The Sound Wave Figure 3 shows what the molecules in a medium look like when a sound wave passes through it. The impulse or disturbance progresses rapidly and travels great distances, but the particles in the medium move only an infinitesimal amount to either side of their rest position. They bump the particles adjacent to them and impart their motion and energy to those particles. In other words, the sound travels, but the medium only vibrates.

The impulse travels in the *same* direction as the movement of the particles—in other words, it is a longitudinal wave motion, not a transverse wave like light.

As can be seen from Figure 2, a *cycle* is a complete single excursion of the molecule. *Frequency* is the number of cycles in a given unit of time, usually cycles per second. (Cycles per second are frequently labeled Hz.)

The *wavelength* of the sound wave is the distance between regions of identical rarefaction or compression.

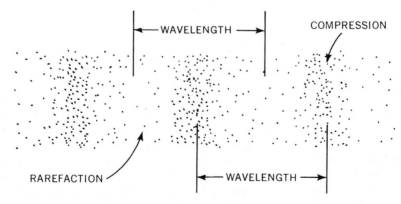

Figure 3 An elastic medium subjected to sound.

The *amplitude* of motion is the maximum displacement, beyond its normal or "rest" position, of the element being considered. In most audible sounds, these excursions are very small, although low-frequency sound may cause large excursions (as would be observed in the motion of a loudspeaker cone reproducing very low frequency sounds at audible level). In some mechanical vibrations, the amplitude of motion can be very great, particularly for very low frequency vibration.

In a given medium, under fixed conditions, sound *velocity* is constant. Therefore, the relationship between velocity, frequency, and wavelength can be expressed by the equation:

$$\text{Velocity} = \text{Frequency} \times \text{Wavelength}$$

In gases and fluids (and in most solids), the impulse expands from the source somewhat like a rapidly expanding soap bubble. The surface of that sphere is called the "wave front." Obviously, then, the energy of the original impulse is diminishing at any one point on the wave front as the square of the distance from the source—the familiar "inverse square law."

Elastic Media The important characteristics of elastic media are mass and density (remembering that mass is the characteristic which imparts inertia to a substance). In addition, elasticity, stiffness, plasticity, and hysteresis are fundamental characteristics of media considered in acoustical design.

Contrary to almost universally held opinion, sound does *not* travel faster in dense media than in less dense media. In fact, sound velocity in media is inversely proportional to the density, as can be seen from the following equation:

$$\text{Velocity} = k\sqrt{\frac{E}{\rho}}$$

where

k = a constant;
E = the modulus of elasticity;
ρ = the density.

What is usually overlooked is the modulus of elasticity of the medium. Usually, very dense materials have very much higher moduli of elasticity than less dense media. As a result, naïve and uncritical examinations of the velocity of sound in various media lead observers to precisely the wrong conclusion. It would seem logical to almost any observer that the progression of the impulse through a medium would be much less restricted in less dense media, and examination of the previous sketches would bear this out. This concept is emphasized because it is fundamental to a true understanding of the transmission or isolation by acoustic media.

TABLE 1 Velocity of Sound in Various Media

Material	Approximate Sound Velocity (ft/sec)	(m/sec)
Air	1,100	335
Wood	11,000	3,350
Water	4,500	1,370
Aluminum	16,000	4,880
Steel	16,000	4,880
Lead	4,000	1,220

Acoustic Force and Energy Concepts The most fundamental concept remains the simple and well-known:

$$Force = Mass \times Acceleration$$

If this were remembered and understood, any intelligent professional could intuitively find his way through many acoustical problems. All materials possess mass; this mass, when moved, particularly when moved back and forth, with change of direction and velocity (which any elastic, oscillating motion must involve) must be accelerated. This process requires force; and force, acting through a distance, is energy.

The amplitude of motion of the particles in the media is dependent upon the sound pressure (force) or the intensity of the sound energy involved.

The energy concepts of interest are kinetic energy, potential energy, energy storage, and energy conversion. Since acoustic phenomena are equivalent to AC electric current, the concepts of resistance, capacitance, impedance, intensity, and pressure are frequently used in discussing acoustical concepts.

The transmission or conversion of the sound energy is the function of so-called "acoustical materials."

Units and Dimensions

The dimension, scale, and proportion of everything associated with these phenomena must be remembered in any discussion of acoustics and acoustical materials. Materials must be considered on the molecular scale as well as the macroscopic scale of panels or constructions. The dimensions of amplitude, wavelength, sound pressure, acoustic energy, etc., must be considered in every problem. If this were done, the rather ridiculous misconceptions often associated with acoustical material design and use would be avoided. "Breaking up" of sound waves with textured paints or effective absorption of sound with thin flocked surfaces would not be considered seriously if we simply remembered the dimensions involved. Even a careful examination of the units of dimension or quantity involved will cast some light on the hard-to-understand concepts involved (note, for example, the discussion of "impedance," Section II, page 59).

The units and dimensions of acoustics continue to be troublesome and puzzling to many people. A simple analysis of the principal units may serve to simplify the problem.

Most of the units of sound measurement are borrowed directly from or are comparable to those of other scientific disciplines. Mechanical, electrical, and acoustical analogies are well-known to all who work in the field of acoustics. The human response to sound energy accounts for many of the units used in acoustics.

Reference Levels For most purposes, the absolute numbers obtained from measurements have little significance in themselves; they are almost always compared with some base or reference, and they are usually quoted as "levels re. . ." that reference. The levels are usually ratios of the reference level, since rarely are simple, linear relationships found between stimuli and effects in humans.

As in many of the physical sciences, the units and reference levels are derived from common, readily observed phenomena. The zero-level of sound pressure, for example, is not a true physical zero (that is, the absence of any pressure in excess of what would have existed with no sound energy present in the medium); rather, it is something of an average "threshold of hearing" at about 1000 cps (Hz) for humans. The physical pressure associated with this level is an incredibly small 0.0002 dyn/cm^2 (often referred to as "microbar," since 1 dyn/cm^2 is about one-millionth of normal standard atmospheric pressure). A sound pressure about one million times as great tends to cause considerable discomfort in humans, with actual pain often accompanying increases above this elevated level. Since intensity varies as the square of the pressure, this range represents a ratio of one trillion to one in energy.

One might assume, then, that a logical procedure would be to divide this range of values into a uniform scale and assign numbers to these steps. However, such uniform steps would have little meaning, since they would have little relationship to human responses. Changes in human response tend to be according to a ratio of the intensity of the stimuli producing the response.

The Decibel If we were to use common units of pressure for the sound pressure scale (lb/in.^{2}, dyn/cm^2, or N/m^2, for example), probably the scale would be more meaningful to everyone, but we would be dealing in enormous numbers which might be difficult to manipulate. Unfortunately, long ago the acousticians borrowed a term from the electrical engineers, and we have lived to regret it.

The "Bel," in electrical measurement, represents a ratio of ten to one. It, too, was chosen to compress a scale of tremendous range. The "deci-Bel," as the name implies, is one-tenth of a Bel. Thus, power levels and intensity levels represent:

$$10 \log_{10} \frac{\text{Quantity measured}}{\text{Reference quantity}} \text{ in decibels}$$

An "Intensity Level," for example, represents the ratio of the intensity being measured to some reference level, and is expressed thus:

$$\underset{\text{(in dB)}}{\text{Intensity Level}} = 10 \log_{10} \frac{\text{Intensity measured}}{\text{Reference intensity}}$$

(Reference Intensity Level = 10^{-16} W/cm^2)

Since intensity varies as the square of the pressure:

$$\underset{\text{(in dB)}}{\text{Intensity Level}} = 10 \log_{10} \frac{(\text{Pressure measured})^2}{(\text{Reference pressure})^2}$$

or

$$= 20 \log_{10} \frac{\text{Pressure measured}}{\text{Reference pressure}}$$

(Reference Pressure Level = 0.0002 dyn/cm^2; 2×10^{-5} N/m^2; or 20 μPa)

So that Sound Pressure Level will correspond with Intensity Level, Sound Pressure Level is defined thus:

$$\underset{\text{(in dB)}}{\text{Sound Pressure Level}} = 20 \log_{10} \frac{\text{Pressure measured}}{\text{Reference pressure}}$$

Sound Power Level refers to the power of a sound source compared with a reference power level of 10^{-12} W. (NOTE: Occasionally, the older 10^{-13} W is used; thus, it is imperative that reference level always be explicitly stated.)

The rationale for this choice of terms was that the ear responds in a roughly logarithmic manner to changes in stimulus intensity. Unfortunately, the correspondence is very rough. In simple, everyday terms, the correspondence is approximately:

TABLE 2 Subjective Effect of Changes in Sound Characteristics

Change in Energy Level	Change in Sound Level	Change in Apparent Loudness
26%	1 dB	Insignificant
Doubling	3 dB	Just perceptible
Tripling	5 dB	Clearly noticeable
Ten Times	10 dB	Twice as loud (or $\frac{1}{2}$)
100 Times	20 dB	Much louder (or quieter)

TABLE 3 Comparison of Intensity, Sound Pressure Level, and Common Sounds

Relative Energy Intensity (units)	Decibels[a]	Loudness
100,000,000,000,000	140	Jet aircraft and artillery fire
10,000,000,000,000	130	Threshold of pain
1,000,000,000,000	120	
100,000,000,000	110	Near elevated train
10,000,000,000	100	Inside propellor plane
1,000,000,000	90	Full symphony or band
100,000,000	80	Inside auto at high speed
10,000,000	70	
1,000,000	60	Conversation, face-to-face
100,000	50	Inside general office
10,000	40	Inside private office
1,000	30	Inside bedroom
100	20	Inside empty theater
10	10	
1	0	Threshold of hearing

[a]SPL as measured on A-weighted network of standard sound level meter.

TABLE 3a Decibel Notation in Acoustics

Power (watts)	Sound Pressure Level dB	N/m^2 (Pa)	Pressure $(lb/in.^2)$	dBA
10^8	200	200,000	3×10	(194 = 1 atmosphere)
10^6	180	20,000	3	
10^4	160	2,000	3×10^{-1}	Jet engines
10^2	140	200	3×10^{-2}	Pain threshold
1	120	20	3×10^{-3}	
10^{-2}	100	2	3×10^{-4}	OSHA limit
10^{-4}	80	0.2	3×10^{-5}	
10^{-6}	60	0.02	3×10^{-6}	Normal speech
10^{-8}	40	0.002	3×10^{-7}	
10^{-10}	20	0.0002	3×10^{-8}	Whisper
10^{-12}	0	0.00002	3×10^{-9}	Hearing threshold

Note: The "decibel" is always a *ratio*, never a unit. In this table, power means watts/m², and dB and dBA are always *levels*.

Decibel Addition Clearly, since they are ratios (logarithmic), decibel levels must not be added arithmetically. Two sources, each producing 50 dB, do *not* combine to produce 100 dB. Rather, they produce 53 dB. (50 dB is a logarithmic quantity. $10 \log_{10}$ of $2 = 3$. Therefore, we add $50 + 3$ to derive the level of the two 50 dB sources.)

TABLE 3b Decibel Addition

Difference Between Two Levels (dB)	Add to Higher Level (dB)
0	3
1	$2\frac{1}{2}$
2	2
3	2
4	$1\frac{1}{2}$
5	1
6	1
7	1
8	$\frac{1}{2}$
9	$\frac{1}{2}$
10	$\frac{1}{2}$
more than 10	0

The Octave Probably borrowing from musical notation, the acoustician frequently used the "octave" in his work. However, he is usually interested in the octave only as a frequency ratio, not as a series of eight intervals. Therefore, in this book the octave will be used only in the sense of a frequency ratio of 2/1.

Thus, the media involved in sound and vibration move twice as fast with each doubling of the frequency—or each octave higher. (NOTE: Velocity of sound remains constant, though.)

2. HEARING

Hearing is the principal subjective response to sound. Within certain limits of frequency and pressure, sound creates a sensation within the auditory equipment of humans and most animals.

At very low frequencies or at very high pressure levels, additional sensations, ranging from pressure in the chest cavity to actual pain in the ears, are experienced.

In most animals and humans, the ear is the receiver. It receives the vibrations on the eardrum, multiplies them by means of small bones arranged as levers in the middle ear, and transmits the vibrations through a fluid to nerve endings within the inner ear. These nerves transmit an impulse to the brain which, in a fraction of a second, analyzes and translates the impulse into a concept which evokes a mental or physical response.

Sounds become, through experience and training, familiar symbols of a concept or situation. They give us information which orients us in our environment.

While the entire mechanism of hearing is still the subject of much argument, there is general agreement that the final step in the process takes place in the brain; the ear is only a receiver. That is why we refer to hearing as "subjective," while sound is "objective." The following sections will show why this distinction is important.

From a strictly mechanical standpoint, the ear responds in a relatively predictable manner to physical changes in sound. The following table relates the objective characteristics of sound to our subjective responses to those characteristics.

TABLE 4 Subjective Responses to Characteristics of Sound

Objective (Sound)	Subjective (Hearing)
Amplitude ⎫ Pressure ⎬ Intensity ⎭	Loudness
Frequency	Pitch
Spectral distribution of energy	Quality

Loudness

Loudness is the physical response to sound pressure and intensity. At any given frequency, the loudness varies directly as the sound pressure and intensity, but not in a simple, straight-line manner.

Humans are very complex animals. Not only are their reactions not linear, but they are nonlinear in an unusual manner. The following graph (Figure 1-4) shows what is commonly called the "envelope of hearing."

The curves are known as "equal loudness contours," and, as the name implies, they represent the Sound Pressure Level necessary at each frequency to produce the same loudness response in the average listener. For example, the lower, dashed curve indicates the so-called "threshold of hearing" (a Sound Pressure Level of 0.0002 dyn/cm^2 —or microbar—is the present 0 dB level). It represents the Sound Pressure Level necessary to produce the sensation of hearing in the average listener. The actual threshold varies as much as ±10 dB among "normal" individuals (and up to +50 dB or so in defective ears). Note, particularly, that we are quite deaf at low frequencies. At 20 cps (Hz), the level must be almost 70 dB higher (ten million times as much energy) as at 2000 cps (Hz) to produce hearing. The upper curve represents the so-called "threshold of feeling," where hearing becomes more akin to pain. Note, also, only at 1000 cps (Hz) are the intervals between curves uniform.

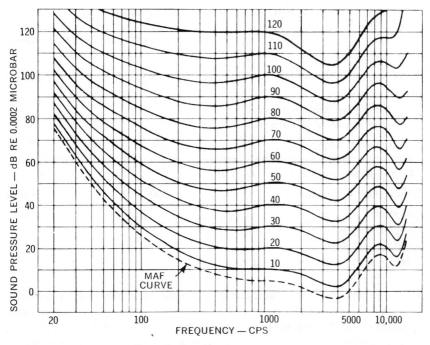

Figure 4 Equal loudness contours.

Pitch

Pitch is the physical response to frequency. Low frequencies are identified as low in "pitch;" high frequencies as "high-pitched." Middle "C," for example, is 261 cps (Hz); one octave below is 130 cps (Hz); one octave above is 522 cps (Hz).

The frequency response of humans is interesting. For practical purposes, our hearing encompasses a range of frequencies from about 16 cps (Hz) to somewhat less than 20,000 cps (Hz). Only at high Sound Pressure Levels is the response reasonably flat throughout the frequency range. This relationship is important to architects and engineers, for it is apparent that some types of sound will affect listeners very much more strongly than will other types.

Sounds generated by mechanical equipment may encompass the entire frequency range of human hearing. Jet aircraft, for example, have significant output throughout the entire range, and large Diesel engines produce substantial energy from 30 cps (Hz) to 10,000 cps (Hz).

The following table lists some of the significant frequency ranges with which the professional should be familiar:

TABLE 5   Significant Frequency Ranges

	Approximate Frequency Range (cps or Hz)		
Range of human hearing	16	to	20,000
Speech intelligibility (containing the frequencies most necessary for understanding speech)	600	to	4,800
Speech privacy range (containing speech sounds which intrude most objectionably into adjacent areas)	250	to	2,500
Typical small table radio	200	to	5,000
Male voice (energy output tends to peak at about:)		350	
Female voice (energy output tends to peak at about:)		700	

It is apparent that low frequencies do not affect us very strongly, while those from 500 to 5000 cps (Hz) (where the ear is most sensitive) are very important. In short, we tend to be very sensitive to sound in the middle and higher frequencies. Loudness, then, is dependent strongly on pitch as well as on the amount of energy in the signal. Our hearing mechanism integrates many of the physical characteristics of sound to evaluate a signal. Further sections of this book will discuss this problem in detail.

Quality Quality of sound refers to the spectral distribution of energy within a signal. Almost every source of sound has its own "voice-print" which identifies the sound with the source. The distribution of energy throughout the audible frequency range gives a distinct, unique character to the sound. Only absolutely pure tones (true sine waves with no harmonics) are identical with other pure tones, regardless of their source. Thus, a violin and a flute, both playing the same note, are quite distinct and recognizable as different from one another. Figure 5 shows the spectral distribution of sound energy generated by a calculator.

Our ability to distinguish different sound spectra and to remember the character of the sounds permits us to identify the source and to evaluate the significance of the sounds. We are particularly sensitive, too, to pure tones; they tend to stand out sharply, even in the presence of high background levels.

Discrimination

Man has an unusual ability to discriminate among the elements of a complex sound and to discern discrete, particular "signals" of interest to him.

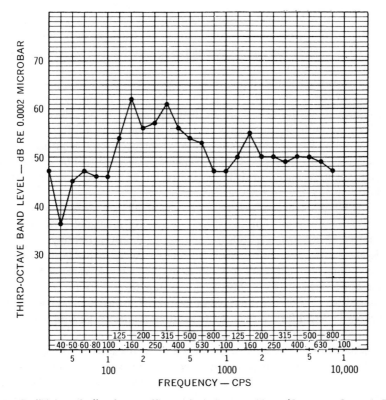

Figure 5 "Voice-print" of an office calculating machine. (Courtesy General Radio Company.)

Various sounds may produce widely different effects in different individuals, depending upon the psychological attitude of the individuals at the time; in fact, identical sounds often produce widely different responses in the same individual at different times.

We live in a "sea of sound," with a constant background of sound of varying level (the minimum would be the "thermal noise" of the constantly moving molecules of the air and the matter which surrounds us). Only when a particular sound or signal becomes sufficiently loud do we detect it over the background. The background, whatever its nature, tends to "mask" sounds until their level exceeds the background.

Signals and Noise   Signal-to-noise ratio is a frequently used term in communication terminology. *Unwanted sound*, whatever its nature, is *noise. Wanted sound*, whatever its nature, communicates *information* to us.

When the "signal-to-noise" ratio becomes sufficiently large, we detect the signal and the information becomes available to us. Thus, the distinction between "noise" and "communication" is completely subjective, completely a matter of our own desires or needs of the moment.

It is imperative that the above concepts be remembered in approaching any acoustical problem. Either the signal (the wanted sound) must be presented at a suitable level above the noise level, or the noise level must be reduced below the signal level—or both. For intelligibility, a signal-to-noise ratio of 10 dB (that is, the signal Sound Pressure Level is 10 dB above the noise level) is sufficient; however, this may not mean that a comfortable or pleasant situation exists. "Outshouting" the noise may provide intelligibility, but at the cost of hearing fatigue, discomfort, annoyance, or even ear damage.

There is a significant level where interesting effects are noticed—a noise level above which even loud, shouting speech is scarcely intelligible. This level—approximately 85 dB in the speech frequencies—is almost universally accepted as very loud or even "uncomfortable." Not unexpectedly, it has been found that continued, long-term exposure to levels in excess of this level tends to produce permanent hearing loss. Even short-term exposure to such levels often produces a temporary increase in the threshold of hearing of normal humans (that is, they are slightly deafened for a while). So, in addition to interfering with speech communication, high-level "noise" can injure the listener.

Depending upon the activity to be carried on in the space, there are levels between the above "uncomfortably loud" and absolute silence (a most intolerable environment for humans, in which they quickly become fearful and disoriented) where humans are reasonably satisfied. These levels are discussed in detail in later sections of this book.

Binaural Hearing   Because we have two ears, we receive signals binaurally, not monaurally as a microphone would. Since each ear receives the signal at a slightly different level and phase (that is, earlier or later in time), each ear probably sends a slightly different signal to the brain. The brain apparently analyzes the signals, integrates them, and extracts information from them, Thus, we can locate the source with a fair degree of accuracy; we can learn something about the type of space in which the sound is created; and we can characterize the source even in the presence of a constant sound background.

If we were to have use of only one ear, we would quickly learn how these skills would be degraded. Listening to a monaural recording of sound just heard "live" in a concert hall, for example, will demonstrate this dramatically.

Human Response to Vibration

In addition to hearing—the human response to "audible" sound—there are other responses to vibration often grouped under the term "feeling." At the risk of some slight technical inexactitude, we will mention briefly these responses.

Actually, the most common objection to vibration in buildings and other normal human environments is the audible effect; the surfaces and components of the environments vibrate strongly enough to turn them into secondary sound sources, often amplifying the original sound source appreciably.

The vibration-induced response called "feeling" is a complex and not entirely understood phenomenon. Current information indicates certain thresholds of feeling as indicated in Figure 6.

Our threshold and tolerance levels vary with frequency (as is true of audible sound, but in quite a different way). There is still little agreement on whether we respond to acceleration, velocity, or displacement, or to a combination of all three.

The circumstances under which people are subjected to vibration appear to influence strongly their response to the vibration. In transportation equipment, for example, we seem to tolerate much more vibration than we will accept in stationary environments.

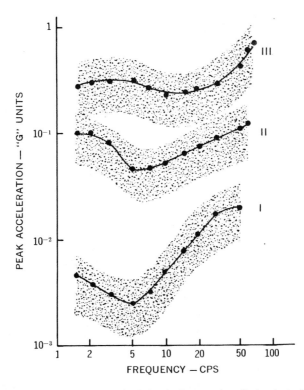

Figure 6 Human response to vibration. I, level of perception; II, level at which vibration becomes unpleasant; III, level at which people refuse to tolerate vibration.)

Physical effects of excessive vibration include disorientation, nausea, damage to tissues, and other phenomena. This aspect of vibration belongs in the realm of medicine and psychology, however; the designer of environments usually finds himself working at vibration levels well below those capable of such dramatic effects.

Section II

3. CONTROLLING SOUND, NOISE, AND VIBRATION

Section I establishes the reason for the remaining sections of this book. If we know what sound is, how it is transmitted, and its effects on man, we can employ scientific and engineering techniques to control it.

Acoustics is the science of sound, including its production, transmission, and effects. Although it is derived from a Greek word meaning "to hear," its meaning has been extended to include sound beyond the limits of hearing.

Acoustics is a broad field which embraces music, radio, sound reproduction, sonar, ultrasonics, and many other specialized subjects of only passing interest to most architects and engineers. The remainder of this book deals with only those aspects of sound control associated primarily with man's normal physical environment—the buildings, sites, and surroundings in which he lives, works, learns, or plays; the equipment associated with such activities; and the environmental factors which affect such activities.

Architectural Sound Control

Occasionally, for brief periods, nature provides an ideal environment for some human activities. The architect or engineer, however, is called upon to provide a full-time, ideal environment for whatever man chooses to do.

A building is, primarily, a controlled environment which facilitates human activity and, as such, influences human conduct and attitudes.

Sound is an integral and vital part of any environment. Thus, to the designer, it is almost like a building material—a plastic medium which can be molded, directed, and manipulated to create the environment that is being sought.

The awe-inspiring hush of a cathedral, the quiet of a library, the brilliance of a concert hall, or a pleasantly quiet factory need not be the results of fortuitous accidents. They can be designed into a structure at will. An acoustical environment can be forecast and designed into any building at the direction of the designer. Acoustics are under his control. How he uses the "science of sound" will determine the end result.

Source-Path-Receiver   In Section I, it was pointed out that sound originates with a source, travels via a path, and reaches a receiver. Thus, sound control may involve any or all of these three elements in the acoustical "equation."

Usually we think of the "path"—the building materials and constructions which contain or transmit the sound—as the domain of the architect. Likewise, we think of the engineer when we think of the "source"; we assume that mechanical equipment is the principal "noise" generator in a space. And, if we remember the "receiver" at all, we may assign him to the safety engineer or the company doctor.

While traditional, this attitude is most unfortunate. It probably explains why we have so many acoustical problems in our modern environments; why "noise pollution" has become such a part of our modern vocabulary.

The fragmentation and specialization of the arts and sciences are particularly unfortunate in the professional activities which deal with man. Man, his activities and his environment are an inseparable whole.

The forces of wind and weather, most of the activities of man, and nearly every piece of equipment man uses are sound sources. All of the matter on which, in which, and with which he lives are sound transmission paths. It is simply not possible to divide these things into neat categories and assign them to one or another of the professions associated with environmental design. The interaction, conscious or otherwise, of all of the elements demands an understanding of all of the factors and a cooperation among all of the people involved.

For these reasons, we have approached the subject of sound, noise, and vibration control as a single, integral problem, rather than a collection of separate topics.

Description, Measurement, and Evaluation   To design an environment requires either the ability to create it with one's own hands, tools, and skills (as did the "master builders" of the Middle Ages, so revered by instructors of architecture) or the ability to describe, define, and specify the characteristics of the environment sufficiently well to permit others to construct it.

We could, for example, specify that a room should be "quiet," or "not noisy," or that it should "sound nice." In fact, some specifications, even today, speak of "soundproof," "absolutely no noise," and other equally absurd requirements. Meaningful description must be much more specific and objective.

Numbers, words, lines and sketches, drawings and specifications are only symbols used to convey concepts. They are objective means to describe subjective requirements or desires. To design, manipulate, or control our environment, we must be able to identify, to measure, and to evaluate the characteristics which are sought.

Measurement involves comparison with recognizable standards; and evaluation implies determining the effect of the measured variables upon performance.

It is relatively simple to measure Sound Pressure Level with a standard sound level meter. Most of the data on sound available to architects and engineers are given in terms of Sound Pressure Level in decibels (often with no reference to the measurement network used, and with no information regarding the frequency range involved). A brief reference to Section I will show how incomplete and meaningless such data can be.

As we may have learned in lighting design, "foot-candles" do not begin to describe a visual environment; they tell nothing of the color of the light, whether it is continuous or intermittent, direct or indirect, diffuse or specular, or a host of other characteristics which determine its visual effect.

Likewise in acoustical design, "dB's" do not describe the acoustic environment.

It is easy today to make a frequency analysis of the sound being measured, and these data are invaluable in truly describing sound. There is no substitute for the "whole curve"—the entire "voice-print" of the sounds being described. However, such descriptions would be impossibly cumbersome to manipulate, and they would be unintelligible to anyone not well-trained in acoustics. A simpler symbology is imperative for design and specification work. We long for simple, one-number specifications to describe even very complex criteria.

Fortunately, acousticians have developed a family of numbers, symbols, "templates and contours" which fairly well describe the more important subjective evaluations of sound and acoustical environments. For most architectural and engineering work, these "descriptions and specifications" are quite adequate; for the rare and special projects, only complete data and the judgment of experienced professionals are sufficient.

Loudness, Noisiness, and Annoyance While it is relatively simple to measure Sound Pressure Level, relating this objective measure to loudness, noisiness, annoyance, and other subjective responses is quite complex.

Fortunately, there is a simple and remarkably practical means of measuring and defining *loudness*, using an old technique and a standard measuring instrument. The sound level meter has three weighting networks built into its circuitry, designed to reflect the built-in "bias" of the ear. The networks are:

"C"-network—which is essentially linear
"B"-network—which reflects the ear's response to sounds of moderate pressure level
"A"-network—which reflects the ear's response to sounds of lower pressure level

Figure 7 indicates the standard weighting characteristics of the modern sound level meter.

Over the years, a wealth of data has been accumulated, using the "A"-network readings of the sound level meter. For most of the normal architectural and

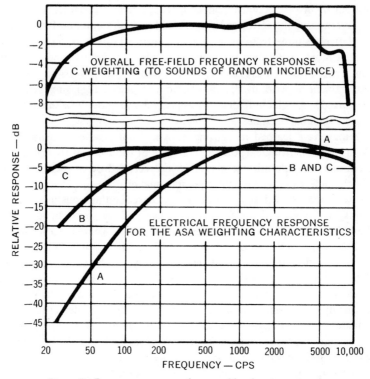

Figure 7   Frequency response for sound level meter networks.

engineering work done today, the "dBA" readings, in decibels, are sufficiently accurate to define the loudness of sounds in the simplest possible way. The readings can be made directly, with no need for additional calculations; and, except for unusual sounds, the values are a good evaluation of relative loudness. Unless specifically stated otherwise, loudness in this book will mean Sound Pressure Level as measured on the "A"-network, in "dBA."

In Section I, page 10, loudness was discussed in detail, and a family of "equal loudness contours" was shown. One of the more frequently used evaluations of loudness involves the term *phon*. If we relate any point on any equal loudness contour to where that contour crosses the 1000 cps (Hz) ordinate, the value at that intersection determines the loudness in "phons" of that sound. For example, a sound of 92 dB at 20 cps (Hz) has a "loudness of 40 phons" (as does 33 dB at 4000 cps [Hz]).

However, 40 phons is not "twice as loud as" 20 phons. The phon scale, being based upon the dB scale, is logarithmic, not linear. Most people feel much more comfortable with a measurement procedure which uses evaluations such as "twice as loud as" or "half as loud as." A procedure for accomplishing this has been devised.

A family of curves, similar to the equal loudness contours, has been drawn, based upon the relative effect of Sound Pressure Level in various frequency bands; and where these curves cross the 1000 cps (Hz) ordinate determines the "Loudness Index" of the sound in *sones*. The relationship between "phons" and "sones" is shown in Figure 8.

Unfortunately, the procedure has been revised frequently during the past several years, and it is still so controversial that its utility is questionable.

Some investigators have found, however, that "loudness" does not necessarily describe *noisiness* or *annoyance*, particularly when comparing the subjective response to sound from various types of aircraft. A method involving "perceived noise level" in "PNdB" and "Noys" (see Figure 9) was developed particularly to compare the "noisiness" of propellor-driven and jet aircraft. These terms are frequently seen in discussions of airport and aircraft noise control, but they will have little immediate applicability for most other work.

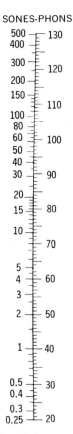

Figure 8 Relationship between sones and phons.

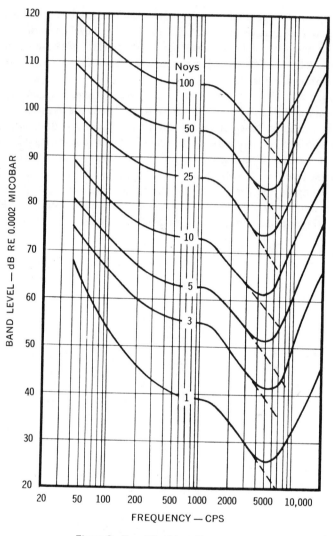

Figure 9 Equal "noisiness" contours.

Noise Sources Noise (unwanted sound) is produced by so many sources that it is safe to say that almost any object or activity—natural, artificial, human, or mechanical—can become a noise-maker. Generally, the sounds which become "noise" include the sounds of communication, transportation, production, recreation, and all of the familiar activities of man. Occasionally, the sounds of wind, weather, water, and even animals become noise. These sounds are an essential part of the environments in which and with which we must operate. (For a listing of the more common noise sources, see Section III, page 94.)

Noise Criteria We cannot tolerate absolute silence for more than very brief periods. In very quiet acoustical environments we become conscious of our own bodily processes, and the slightest movements and even insignificant sounds become disturbing. We almost always prefer some "masking sound"—a form of "acoustical perfume"—to cover up the little sounds which might otherwise become annoying or distracting.

Sounds loud enough to interfere with sleep or with listening tasks are usually undesirable. However, *privacy*—rather than silence—is what we usually want in

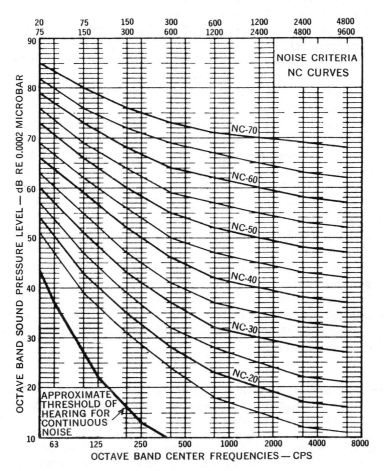

Figure 10 Noise criteria curves.

an environment. We want to be protected against the intrusion of stimuli which interfere with our task or mood. Unless very loud, extraneous speech sounds do not particularly disturb us during our waking hours until they become distinct enough to carry information—that is, until they are intelligible. Likewise, we rarely worry about our own conversation being transmitted to other areas unless it has some private informational content which we do not wish to share.

Therefore, some background sound is usually not only acceptable but desirable.

On the basis of extensive research into human response and preferences, acousticians have developed a set of contours, very similar to the equal loudness contours, to describe numerically various levels of acceptability of steady, constant background sound. These contours (Noise Criteria Curves) are a sort of template which specify the maximum level in any frequency band which will be acceptable in the environment being specified.

When an "NC-number" is used to specify an environment, it means that in no frequency band shall the Sound Pressure Level in the space (measured in the region of normal human activity within the space) exceed the specified NC-Curve (indicated by the NC-number). Usually this type of specification is quite satisfactory for normal, broadband sound, containing no pure tones or sharp, discrete "peaks."

Most people agree that environments with background sound levels under NC-30 are "quiet"; those with levels above NC-55 are "noisy."

For various activities, various background levels are acceptable. Radio and recording studios, for example, can tolerate little background noise; the microphones will pick up the noise and the tapes will record it. In such spaces, NC-20 is a common maximum. In large, general office areas, levels as high as NC-50 are tolerated without much complaint.

(Section III, page 88, lists typical background levels for various typical spaces and neighborhoods.)

We have avoided calling background sound "noise" unless and until it *exceeds* the specified criterion. Noise is *unwanted* sound; sound below the specified criterion is *wanted*, and cannot be properly described as noise.

With these measurement tools and specification criteria, the architect and engineer are equipped to measure, describe, and specify the acoustic environments with which they must deal.

(**Note:** Not surprisingly, there is a remarkable similarity between NC-numbers and dBA values. For most normal, usual background sound, the NC-number is 7 to 10 points lower than the dBA readings for that sound. Therefore, in the absence of a complete octave-band frequency analysis of a background, it is safe to assume that the dBA measurement with the standard sound level meter may be reduced about 7 to 10 points to give the NC-number for the environment.)

Determining the Acoustical Environment The initial step in architectural acoustics is to determine the acoustical environment for the spaces being designed. In order of criticality, spaces can be defined subjectively as:

Quiet — where a minimum of background sound is acceptable. Radio studios, sleeping rooms, etc., fall into this group.

Hearing spaces — where listening attentively to the source is the principal activity within the room. Theaters, concert halls, churches, lecture halls, and the like, are typical.

Normal — where communication, normal "not noisy" human activities are conducted.

Noise — where a general impression of "loudness" or elevated sound levels is immediately apparent when entering the room.

Very Noisy — factories, airports, railroad stations, and other environments in which conversation often takes place at elevated levels, and humans are glad to leave the space for relief from the "noise."

(A detailed tabulation of the more common spaces and their Noise Criteria is given in Section III, page 88.)

The *site* of a building and its location on the site are equally critical in acoustical design.

Whenever possible, choose a site consistent with the use of the building. Don't put a rest home next to a railroad classification yard, nor a concert hall adjacent to a drop-forge plant. Absurd as these things may sound, such mistakes are made much too frequently to be ignored.

Occasionally, for various reasons, a critical building must be located on a difficult site. There are solutions to such problems, but they are complex and expensive. It is possible to solve almost any acoustical problem—if the client can afford it. Often, however, it is wiser to choose another site than to try to shut out the sound and vibration present in some areas. (Typical background levels for various types of neighborhoods are listed in Section III, page 89.)

Location of the building on the site is also important. Whenever possible, locate the building as far from the noise sources as possible. Use existing shielding or barriers, such as hills, embankments, other buildings, and the like, to shield the building and to minimize exposure to the noise.

Trees and vegetation are *not* particularly useful sound barriers. Even a dense planting, at least 100 ft (30 m) deep, will provide only 5 to 8 dB shielding. Deciduous trees provide almost no shielding during the months when their leaves have fallen. Grass and shrubbery are useful principally for appearance—not for noise reduction.

Orientation of the building on the site is a particularly useful means of reducing noise problems. Use the less critical areas of the building to shield the more critical from direct exposure to the noise. Place the playground between the expressway and the school building (we assume that a high, strong fence will keep the youngsters out of the road). Locate the loading docks on the noisy exposure, and put the private offices on the quiet side.

Today it is very important to remember the overhead exposure, too. Check for approach and takeoff patterns for existing and planned airports.

(See Section III, page 91, for suggestions on site planning.)

Layout and Arrangement of Spaces   The next planning step includes the layout and arrangement of interior spaces within the building. Until we have determined the difference in level between spaces or between the inside and outside of the building, it is not possible to establish the isolation requirements for partitions or for the exterior construction of the building.

Generally, it is best to group spaces according to their internal sound level— quiet spaces in one area, noisy spaces in another, and "buffer" spaces between them. Corridors, storerooms, closets, and similar spaces can be used to separate areas and to minimize the difference in sound level between them. Usually it is best to put the noisier spaces on the noisier outside exposures, the quiet spaces on the quieter exposures. Also, less critical spaces can be used to surround (horizontally and vertically) the more critical spaces.

Obviously, it is unwise to put the President's Office under the penthouse Machine Room—yet this is one of the more common errors in building design.

Decisions such as these must be made knowingly, carefully, and as early in the design process as possible.

(See Section III, page 93, for additional discussion of this aspect of acoustical planning.)

Choosing the Exterior Construction   Now—not before—the designer is in a position to choose the materials and construction of the exterior of the building. The exterior—including the roof, fenestration, louvers, and all similar significant "details"—must be designed to provide an adequate barrier between the outside "noise" and the interior of each space.

Since the outside sound levels can be measured or estimated, and the acceptable inside levels can be determined, the isolation to be provided by the exterior of the building can be determined readily.

Occasionally, as in the case of industrial buildings, the problem is to keep the noise *in.* The surroundings may be quieter than the activity within the plant. However, the problem is essentially the same; the exterior of the building must function in the same way, only the direction has been changed.

It is apparent that lightweight curtain wall, the huge expanse of plate glass, or the big, attractive louvers rather than heavy doors may *not* be such a good idea if they face noisy exposures. It is folly to design the exterior elevations before determining the extent of the acoustical problems associated with the particular building and its site.

(For further discussion of the function and performance of sound barriers, see "Acoustical Materials," Section II, page 32. For performance criteria and ratings of various constructions, see Section III, page 98.)

Choosing the Interior Construction Everything used to construct and furnish the interior of the building must be selected for its acoustical characteristics as well as for its ability to perform other functions in the building.

The first choice involves the ability of the materials, systems, and constructions to contain or "isolate" sound; they must be acoustical barriers to prevent transmission of undesirable sound, impact, or vibration, whether airborne or structure-borne.

The next decision probably involves the ability of the materials to reflect or absorb sound.

Subsequent decisions include the effect of the constructions on diffusion, focusing, diffraction, attenuation, and reinforcing of sound within spaces.

Surface materials, too—floor, wall, and ceiling coverings—are as much "acoustical materials" as they are wearing surfaces, fireproofing, or decoration.

In short, all interior materials and constructions affect the acoustical environment and they influence the character of the sound in the spaces in which they are used.

Specific characteristics of these elements of interior building construction are discussed in detail in subsequent sections of this book; however, it is imperative that the designer not ignore the implications of decisions made at this stage of design. Acoustical considerations are fundamental—not peripheral or superficial.

(Refer also to Section II, page 31, "Acoustical Materials"; and Section III, page 98, for performance criteria and ratings of materials and constructions.)

The Shape and Configuration of Spaces The shape, dimensions, and proportions of a space are major determinants of the acoustics of the space. Rooms are much like musical instruments—volumes of air within specially shaped enclosures. The surfaces which bound a space affect and control the sound within the space.

For spaces in which critical listening is the major purpose of the room, the shape should evolve from acoustical requirements, not from preconceived prejudice or stylistic whim.

For example, an auditorium (literally, "a space for listening") must be designed in every respect to fulfill its acoustical function; all other considerations, however important, are secondary.

The room surfaces which reflect sound may either concentrate and focus the sound or diffuse and disperse it.

Since it takes a finite and appreciable time for sound to travel any distance, the dimensions of a room (distance between surfaces) and the orientation of the surfaces with respect to one another are significant. Reflections of sound from surfaces may be either reinforcing and helpful or interfering and harmful. Usually, reflections which arrive at the listener's ear within 0.04 sec of the direct sound are desirable (if they come from the right direction, not too different from the source direction); they reinforce and enhance the direct signal. However, reflections arriving later than this may cause undesirable effects.

A major problem in almost any space is that of *echoes*—reflections which interfere with good hearing. The three types of echoes which are particularly important to the designer are:

1. Distinct, discrete sounds, arriving at the listener's ear sufficiently later than the direct sound to be heard as a delayed image of the direct sound (usually 0.06 sec or more—the time required for sound to travel about 70 ft [21 m] in air).

Thus, any reflecting surface more than 35 ft (11 m) from the source and facing the source may produce echoes.

2. "Flutter" echoes—the "rattle" or "buzz" sound resulting from discrete, rapid, multiple reflections between closely spaced parallel surfaces, but too rapid to be readily distinguished as images of the direct sound.

Usually surfaces must be more than 15 to 20 ft (5 to 6 m) apart to cause flutter, although long or high corridors and similar spaces may produce serious flutter with less distance between surfaces (between walls or between floor and ceiling).

3. Reverberation—the persistence of sound after the source has ceased—resulting from the blending of many reflections into an indistinct but apparent "sea of sound." Reverberation may be desirable (as in the case of music), but it may seriously reduce the intelligibility of speech if it persists too long.

The "build-up" of noise within a space is precisely that—an increasing level of acoustical energy as the source continues to emit sound, until the losses (from transmission or absorption) equal the input of the source. When the source ceases, the sound "dies down" as the energy is dissipated.

Sound-absorbing materials are used to reduce the build-up (a process often called "noise reduction") and to reduce the intensity of each reflection so that the reflections do not interfere with hearing the direct signal clearly. This form of reverberation control is frequently overemphasized in books on architectural sound control to where many professionals feel that an "acoustical analysis" means principally the calculation of the reverberation time within space. Actually, in a well-designed space, with proper attention to other acoustical problems in the space, proper reverberation time results almost automatically.

There is a range of "optimum" reverberation times for rooms of various sizes used for various purposes. Except for exotic "dead" (anechoic) rooms and very reverberant test chambers, the optimim reverberation time varies directly with the volume of the space, and ranges from about 0.5 sec for small rooms to more than 2 sec for large spaces. Usually, rooms designed primarily for speech require shorter reverberation times than those for music. There is something of a mystique associated with this subject, much of it rooted in tradition, habit, unexplainable preferences, and taste. Speech intelligibility normally does require reasonably short reverberation times—usually under 1.5 sec in any space, and preferably nearer 1 sec or less for lectures, drama, and motion pictures.

Older music, particularly organ and symphonic music, was composed for the large, reverberant halls in which it was performed. Hence, a somewhat reverberant space is more hospitable to the performance of such music.

(For additional discussion of reverberation time criteria, see Section III, page 135. The use of acoustical absorbents to control reverberation, and formulas for calculating reverberation time and noise reduction are discussed in Section II, page 57.)

According to their shapes, surfaces focus and concentrate or diffuse and disperse sound which reaches them. The wavelength of sound is so much greater than that of light, that the scale of the sound reflector is much larger than the optical reflector. Until a protuberance or indentation approaches 4 to 6 in, (10 to 15 cm) in depth, it affects only the very high frequencies of sound, beyond our normal range of interest. Low-frequency reflectors must be of large area (dimensions of the order of 48 in [1.2 m] or more) and considerable depth (well over 6 in [15 cm]).

Usually, a reasonably diffuse sound field is preferable to a space with highly specular reflections. Usually, also, convex surfaces are much safer than concave surfaces in almost any space. Concave surfaces can focus and concentrate sound undesirably, causing "hot spots" of excessive loudness and other "dead spots" with undesirably low level.

Some types of surfaces may cause undesirable *diffraction* effects. These effects may be an undesirable absorption of energy from important frequency ranges (notably 250 cps [Hz], near middle "C") or generation of peculiar "squeaks" or almost pure high-frequency tones. Particularly dangerous in this respect are uniformly spaced slats, small "flying" panels, rows of seat backs, and other unintentional "diffraction gratings." In general, such designs should be avoided whenever possible.

A reasonably uniform distribution of sound energy within a space is usually desirable. A slight "fall-off" in level from the front (sound source) to the rear is not objectionable if the reduction is small and uniform rather than large and "spotty."

Obviously, a small space presents fewer problems of level and distribution than a larger space. Even a relatively weak source can "fill" a small space with adequate acoustic energy, and reflections are many and frequent in such a space. In a large space, even a strong source may not be able to provide adequate level in remote areas without assistance. A really fine legitimate theater, for example, will have less than 1000 seats (about 600 to 800 is even better); a concert hall should have less than 3000 seats (2200 to 2400 is better); and a lecture hall for unassisted voice should have less than 300 seats. Electrical amplification is usu- ally required for such spaces with larger capacities, and this presents another new set of problems (discussed later, in Section II, page 67, "Sound Amplification Systems"). The tendency toward larger and larger performing spaces (dictated by economic pressures in most cases) has created many acoustical problems, most of them insoluble without sound amplification or reinforcement systems.

In summary, any surface which does not absorb a major portion of the acous- tic energy which reaches it will become a sound reflector. It is automatically a significant element in the design of the space in which it occurs.

The "Sending End" and the "Receiving End"   The source or "sending" end of a space should normally be somewhat "hard" and reflective to project the en- ergy out into the listening or "receiving" end of the space. The receiving end should not reflect much energy back toward the source (the speaker or musicians or performers), or the source may receive a distinct and annoying echo.

For small rooms in which the direct sound reaches the listener at a reasonably high level, this is not vital; but for large rooms it is very important. The source energy must be conserved and directed out into the audience.

In concert halls and music spaces, particularly, it is essential that the musicians be surrounded by hard, reflective diffusing surfaces, not only so that their sound will be sent out into the audience, but so that they can hear one another, and their sounds will blend and be in balance. Even rehearsal rooms (band and or- chestra practice rooms, choral rooms, etc.) should be designed on this basis, with the musicians surrounded by reflective surfaces and facing absorptive surfaces behind the conductor (to simulate the absorption of the audience and seating area of a real performing space).

In short, the sending end of a space should be hard and reflective, and it should "look into" a more soft, absorbent receiving end of the space.

(See Section III, page 137, for further suggestions on shaping spaces.)

Control of Impact, Vibration, and Structure-borne Noise   Airborne sound can be controlled by means of absorbents, reflectors, and barriers; but *impact* and structure-borne noise and *vibration* must be controlled by other means.

Even very thick and massive solid barriers conduct impact sounds quite readily. For example, feminine spike heels on a heavy, unprotected concrete floor can be heard distinctly in the space below. The vibration of a reciprocating compressor, bolted rigidly to an upper floor of a building, may disturb occupants hundreds of feet away. A small not particularly noisy office machine may set a tabletop into resonance, and the tabletop will become an effective sound radiator.

Controlling structure-borne noise and vibration requires:

1. Preventing acoustical energy from getting into the structure. This involves the use of resilient floor coverings such as carpet, resilient mounting systems and supports for vibrating equipment, and the like.
2. Interrupting the transmission paths. This involves the use of construction systems which employ resilient connections and supports for various elements of the systems, avoiding long, continuous, rigid transmission paths.
3. Choosing construction systems which minimize energy transmission. This involves avoiding light, continuous, uninterrupted diaphragm constructions which, by means of shear waves (like the shaking of a rope) transmit energy for long distances. Balloon framing, light curtain walls, thin concrete on metal forms on bar joists, and similar constructions are examples of systems susceptible to problems.
4. Minimizing radiation from surfaces. This involves damping vibrating surfaces by changing their configuration or applying damping materials.

(For further discussion of these subjects, see Section II, "Acoustical Materials," below; and Section III, page 157; also "Mechanical Equipment Noise and Vibration Control," Section III, page 153.)

Acoustical Materials

It is probably unfortunate that the term "acoustical materials" has become a part of our technical vocabulary, since this suggests a family of unique, specific materials with unique properties. Even the architects' standard contract documents relegate acoustics to Section 9 of the specifications and to the finish schedule on the working drawings.

Actually, *all* materials are "acoustical materials" in the strict sense of the term. They absorb, reflect, or radiate sound, and they damp vibrations. The acoustical characteristics of all materials are as basic as their density, elasticity, or hardness. In fact, the acoustical characteristics of materials are directly related to the basic physical properties of the materials; this should be understood and taken into account when a material is being considered during design.

While technological developments tend to make obsolete particular products and materials, the basic characteristics which make materials "acoustical materials" are easily defined and identified. New products and materials will continue to be judged and selected according to the well-established parameters which will be discussed in this section.

During the past few decades, a whole fabric of myths, partial truths, preconceptions, and misconceptions about acoustics and acoustical materials has developed, even among the technically trained. A discussion of the simple, basic principles follows in order to equip practicing professionals, at least, with a fundamental knowledge of these subjects.

Acoustical materials are essentially transducers. Usually, they convert some of the sound energy which reaches them to thermal energy. The reflection, transmission, radiation, and absorption of acoustic energy by various materials constitutes essentially the whole field of sound and vibration control.

Sound Barriers   The most important acoustical materials are those which reflect, contain, or "isolate" sound.

The designer usually is interested in barriers to contain acoustic energy or to block transmission of airborne sound from one space to another. Containing sound, providing a barrier against its transmission via the air, is undoubtedly the major problem in most sound control work.

It is well-known that a wall or heavy enclosure will serve as a very effective barrier against airborne sound transmission. While any surface will reflect some of the sound which reaches it, only heavy, airtight surfaces are significantly effective in containing or "stopping" sound. The more massive and airtight, the more effective sound barriers they become.

In spite of a tendency to speak of constructions as "acoustical" or "soundproof," their function as sound barriers is usually only to provide a means of maintaining a certain difference in sound level between spaces. Their requirements and their effectiveness in this function can be determined and specified. Figure 11 shows graphically what is meant by this.

If a higher level, say 60 dB, exists on one side of a wall, for example, while an acceptable level on the opposite side is not more than 30 dB, the wall must provide at least 30 dB isolation to keep out the intruding sound. If the wall provides 40 dB isolation, the intruding sound will be at a level of 20 dB—safely below the acceptable level in the quieter room.

Thus, a sound barrier, be it wall, floor, or partition, need provide only enough isolation to keep intruding sound below the desired level in the space which it is

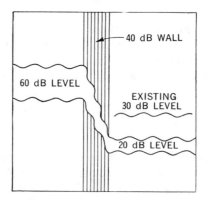

Figure 11   A wall as a "sound barrier."

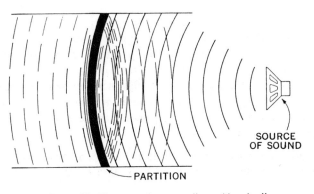

Figure 12 How sound moves a "sound barrier."

protecting. (See Section III, page 98, for isolation criteria for barriers between spaces.)

The Mechanism of Sound Isolation When a wave front reaches a barrier, the barrier is set into motion. The barrier, then, becomes a sound source, and sets into motion the air on the other side. Figure 12 shows how this occurs. Some of the energy is transmitted to the air on the opposite side of the barrier; some of the energy is reflected back toward the source; and some is lost in moving the partition.

Even a thick, massive barrier will move slightly. However, as we know instinctively, the motion will be small, and it will take considerable force to move a heavy barrier. But even a tiny hole will offer a path of low resistance through which sound transmission will occur readily.

Since the barrier moves with an oscillating and accelerated motion, it obviously requires force to initiate and sustain the motion. The partition has mass; it is accelerated by the force or pressure of the impinging sound wave. Therefore, it is possible to analyze its motion mathematically.

From the laws of motion, we know that:

$$\text{Force} = \text{Mass} \times \text{Acceleration}.$$

The instantaneous kinetic energy of the moving partition is proportional to:

$$\tfrac{1}{2} MV^2$$

where

$$M = \text{the mass};$$
$$V = \text{the velocity}.$$

Therefore, we know that more force (pressure) and more energy are required to vibrate a panel at higher frequencies than at lower frequencies. With each octave increase in frequency, the sound energy increases four times (in proportion to the square of the velocity) for a given panel mass.

For each doubling of the mass of the partition, the pressure (force) must increase two times to maintain the same motion; and, since energy is proportional to the square of the pressure, the energy increases by four times.

Thus, the energy expenditure to maintain a given amplitude of motion for a panel increases by 6 dB per octave frequency increase and 6 dB per doubling of the mass per unit area of the panel.

(**Note:** Relative levels in dB are expressed as $10 \log_{10}$ of the ratio of values. Thus, $10 \log_{10} 4 =$ approximately 6 dB. Refer to Section I, "Units and Dimensions.")

Sound Transmission Loss The ratio of the sound energy incident upon one surface of a partition to the energy radiated from the opposite surface is called the "Sound Transmission Loss" of the partition. The actual energy "loss" is partially reflected energy (back toward the source) and partially heat (internal losses within the partition).

The term "loss" may be confusing, since it suggests that the barrier loses or "leaks" the energy. Usually, quite the opposite is true. The signal level at the noisy face of the barrier is higher than the level at the opposite (quiet) face; this difference is due essentially to the ability of the barrier to *prevent* transmission through it. The "loss" is in the strength of the signal radiated from the quiet surface of the barrier. Thus, a "high Transmission Loss" barrier is a good one; a "low Transmission Loss" indicates a poor barrier. Perhaps "Sound Isolation" would be a better term, since "high Sound Isolation" would correctly describe a barrier's performance; but "Sound Transmission Loss" is too well-established to be displaced, so it will be used throughout this book.

Sound Transmission Loss is an inherent characteristic of a barrier and is essentially independent of the location of the barrier.

(**Note:** We do not actually "hear" the Sound Transmission Loss, nor can we measure it directly. We can hear and measure the difference in Sound Pressure Level between two spaces separated by a barrier, and we call this difference "Noise Reduction." It includes the effect of the absorption present in the receiving room—the room with the lower level—and any other energy losses which may be occurring.

In test laboratories, following ASTM Test Method E90-75, a test panel covers an opening between two rooms constructed with thick, massive walls which transmit much less energy than the test panel [therefore, essentially all of the transmission between rooms can be considered to take place through the test panel] ; and the absorption in the receiving room is known. A sound source is located in one room, and the Sound Transmission Loss is determined thus:

$$NR = SPL_s - SPL_r$$
$$STL = NR + 10 \log_{10} S/A$$

where

NR $\quad$ = Noise Reduction;

STL $\quad$ = Sound Transmission Loss;

SPL_s = Sound Pressure Level in Source Room;

SPL_r = Sound Pressure Level in Receiving Room;

S $\qquad$ = Area of Test Panel;

A $\qquad$ = Total Absorption in Receiving Room in units consistent with S.

As is apparent, when little absorption is present in the Receiving Room, the STL exceeds the NR; and when much absorption is present, the opposite may be true.)

Limp Mass Law If the barrier were a "limp" mass, and moved only back and forth (like the end of a piston), the Sound Transmission Loss for energy randomly incident on the barrier (excluding losses at the edge of the panel and any "leaks") would be calculated as:

$$TL = 20 \log_{10} W + 20 \log_{10} F - 33$$

where

W $\quad$ = weight in lb/ft^2;

F $\quad$ = frequency in cps (Hz);

TL = Transmission Loss in dB.

or:

$$TL = 20 \log_{10} W + 20 \log_{10} F - 47$$

where

W $\quad$ = weight in kg/m^2;

F $\quad$ = frequency in cps (Hz);

TL = Transmission Loss in dB.

Coincidence Effect Rarely, however, does a panel act as a limp mass. Usually it moves in a more complex manner, depending upon its *stiffness*. Often a significant "shear wave" (comparable with the transverse waves created in a vibrating string) occurs in the panel. When the velocity of this shear wave coincides with the component of velocity of the incident sound wave in the air (usually sound impinges on the surface from all directions, not just at right angles to the wall), Sound Transmission Loss of the panel is sharply reduced. Theoretically it would drop to zero, but internal losses within the panel provide appreciable attenuation. Actually, a "plateau" occurs in the Sound Transmission Loss curve, quite different from the simple 6 dB/octave "limp mass" curve. (See Figure 13.)

Thus, the performance of a panel of a given material varies not only with the surface mass but with the elasticity or stiffness of the panel. Lightweight, stiff

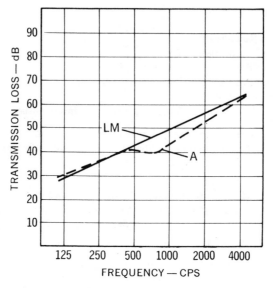

Figure 13 Comparison between Transmission Loss of "limp mass" and actual panel. (LM, limp mass; A, actual solid panel.)

panels (light, "honeycomb" core panels, for example) tend to behave much more poorly than their mass alone would indicate. Dense, "limp" materials, such as soft lead, behave nearly according to the mass law throughout much of the frequency range.

Double-wall Construction   It is apparent that two serious natural limitations exist in all real materials which might be considered for sound barriers:

1. Truly "limp" materials have little use in the construction of most enclosures.
2. If a doubling of mass produces only about 6 dB improvement in Sound Transmission Loss, a panel becomes prohibitively heavy when very high Transmission Loss is required.

Fortunately, a given mass of material may be used in a way that appreciably improves its Sound Transmission Loss throughout most of the significant frequency range. If the mass is divided into separate layers with no rigid connections between layers, a substantial increase in performance occurs (see Figure 14).

The layer of air between layers of surface material (unless very thin) is "limp" enough to provide poor energy transfer from one surface to the other. The shear wave in one surface is only very inefficiently coupled to the opposite surface; and the pistonlike motion of the surfaces is "cushioned" by the soft layer of air

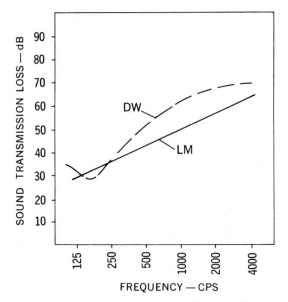

Figure 14 Comparison between Transmission Loss of "limp mass" and same mass divided into double wall. (LM, limp mass; DW, double wall.)

between layers. The impedance match between the surfaces and the entrapped air is very poor except at certain resonant frequencies, and even at those frequencies, internal damping provides substantial attenuation.

If an absorbent blanket is placed in the air space between layers, the air must "pump" back and forth through the absorbent material, further attenuating the energy.

Theoretically, each doubling of the air space between surface skins of a double-wall partition should improve Sound Transmission Loss by about 6 dB. Actually, the gain is somewhat less than this in actual constructions—more like 5 dB or less.

Where structural requirements preclude total separation between surface layers, a relatively soft inner layer may be used. Its effectiveness depends upon its shear modulus. Very low shear modulus materials for the inner layer permit each panel to move in shear, somewhat independently of the opposite panel. As a result, the barrier tends to approach its "limp mass" Sound Transmission Loss potential.

Sound Transmission Class As is apparent from Figures 13 and 14, only the complete Transmission Loss curve provides a complete description of the performance of a barrier in isolating against airborne sound. However, a listing of 16 values at 16 frequencies would be required simply to specify the Transmission Loss of a wall or partition.

Understandably, architects and engineers have a strong predilection toward single-number ratings for materials and constructions. Therefore, a scheme has been devised for comparing the actual Sound Transmission Loss curve against a "standard contour" (related to the importance of isolation required at various frequencies; comparable with the shape of the "A-weighting" curve of the sound level meter, and approximately the inverse of the equal loudness contours; see Section II, "Loudness," page 19).

The contour is "fitted" to the actual test performance curve, and the Transmission Loss value where the contour intersects the 500 cps (Hz) ordinate is called the "Sound Transmission Class" (STC) of the construction. (See ASTM E 413-73 "Determination of Sound Transmission Class.")

Note in Figure 15 how two partitions with identical "averages" for their Transmission Loss can have STC ratings which differ by as much as 10 points. The deep "dip" in the curve for the STC-22 partition is a serious deficiency in its isolation performance, yet the "average" of values at the various frequencies

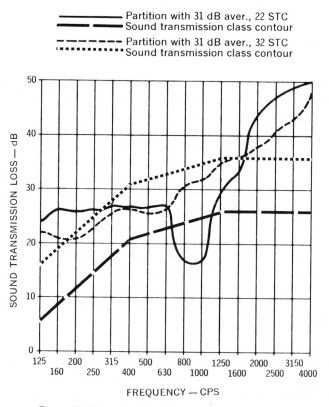

Figure 15 Determination of Sound Transmission Class.

disguises the deficiency. The STC-32 partition is very much better as a sound barrier, as the rating indicates.

It is possible to forecast roughly the Transmission Loss to be expected from a panel, knowing its mass alone. However, as discussed previously, the coincidence dip, structural separation, and construction details affect performance significantly. Most manufacturers of materials and constructions, and various agencies and trade associations, publish current Transmission Loss data.

The range and magnitude of Sound Transmission Loss performance for various materials may be inferred from the following table of typical panels used in building and product design. (See Section III, page 105, for a detailed listing of the STC-ratings for most of the standard wall, partition, and floor construction systems used in the building industry in the United States.) It is important to remember that the performance of a barrier is strongly dependent upon numerous details of construction, possible "leaks" and "flanking" (a discussion of these problems follows), and the control of many variables associated with the design and construction of the barrier. The values shown in this book are often somewhat lower and more conservative than those shown in other published literature and advertising brochures. A laboratory test, properly conducted, indicates what a relatively "perfect" installation is capable of providing; it is probably the most "accurate" rating for the construction in most cases, but it is rarely the realistic rating which a designer should use. The values shown in this book are based upon an analysis of more than 7000 tests, weighted heavily with field installation tests. They represent a realistic evaluation for the constructions in careful field installations with no major "leaks" or significant "flanking" losses.

Impact Insulation In addition to serving as barriers against airborne sound, floors must provide protection against impact noise. The sounds of walking,

TABLE 6 Sound Transmission Loss for Common Building Materials

		Material	*STC-rating*
$\frac{1}{4}''$	(6 mm)	Steel plate	36
$\frac{1}{4}''$	(6 mm)	Plate glass	26
$\frac{3}{4}''$	(19 mm)	Plywood	28
$4''$	(10 cm)	Brick wall	41
$6''$	(15 cm)	Concrete block wall	42
$\frac{1}{2}''$	(12 mm)	Gypsum board on both sides of $2'' \times 4''$	
		(5 cm × 10 cm) studs	33
$12''$	(30 cm)	Reinforced concrete wall	56
$14''$	(35 cm)	Cavity wall consisting of	
$8''$	(20 cm)	Brick–$2''$ (5 cm) air space–$4''$ (10 cm) brick	65

moving furniture, dropped objects, etc., become noise if they intrude objection-ably into adjacent areas.

The airborne sound isolation provided by a floor bears little relation to its impact isolation. In fact, the whole mechanism of impact sound transmission is quite complex (and is related to the "impedance" of the construction; see Section II, page 59, for further discussion of this subject).

(**Note:** "Floors," in this section, refers to "floor/ceiling assemblies." As will be seen in Section III, page 114, the ceilings attached to or suspended from floors appreciably affect the airborne and impact isolation of the construction.)

Impact Noise Rating Systems   Unfortunately, at this time no satisfactory test method exists for evaluating the impact isolation of floor constructions.

Some U.S. Government agencies (FHA and HUD, for example) recommend and publish ratings based upon an International Standards Organization test method which uses a tapping machine of known performance characteristics as an impact "source." The machine "hammers" on the floor specimen, and the Sound Pressure Level is measured in the room below. Therefore, the better the floor, the *lower* the level in the receiving room.

Again, a "standard contour" which reflects subjective response to noise (but approximately the obverse of the STC-contour) is fitted to the curve of the level measured in the receiving room. The relative vertical position of the contour determines the "Impact Noise Rating" of the floor; and it is said to compare and rank-order constructions.

Figure 16 shows the "Standard Contour" used by FHA, compared with those used in other countries.

The older FHA procedure identified the standard contour as an acceptable level in multifamily dwellings, and rated the level indicated in Figure 16 as "0." Constructions producing levels *below* the standard shown were rated as "+," indicating superiority over the standard; those producing levels *above* the standard were rated as "−," indicating inferiority. This confusing situation has been clarified somewhat by the newly proposed HUD procedure which identifies the older FHA "0" as approximately "51" on its rating scale, thus producing positive numbers for almost all constructions. Older "INR" (Impact Noise Rating) figures can be converted to the newer "IIC" (Impact Insulation Class) ratings by adding 51 algebraically to the INR values.

The tapping test method is based upon experience gained largely from tests of the heavy masonry floor constructions widely used in European apartment and multifamily buildings. It tends to overrate the effectiveness of floor coverings such as carpet; it does not appear to evaluate properly the effect of heavy, slow impacts or of low-frequency impact sound; and it does not reflect accu-rately the performance of lightweight floor constructions such as the more common U.S. wood-joist floors and similar systems.

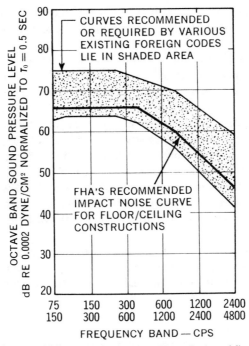

Figure 16 Determination of Impact Noise Rating of floors.

Performance criteria for floors in various occupancies have not been agreed upon very widely, and floors with widely different performances are apparently providing acceptable isolation in actual installations. Generally, a floor rated "INR-0" (approximately "IIC-51") is acceptable for most occupancies. A reasonably stiff floor, weighing not less than 30 lb/ft² (150 kg/m²), and covered with carpet on a good pad is usually adequate. However, additional construction details and specifications must also be considered in designing acoustically acceptable floors. Experienced architects and acoustical consultants usually draw heavily upon their experience to evaluate constructions, regardless of published data (or the absence of data).

"Leaks" and "Flanking" The sound Transmission Loss rating of a barrier is based upon the assumption that all of the energy incident upon one face of the barrier is radiated from the opposite face only. In practice, however, this may not be the case. Frequently, appreciable transmission occurs via "leaks" and "flanking" paths.

Leaks refers to holes and openings through the barrier or at its perimeter; airborne sound is transmitted readily via such paths.

Flanking usually refers to structure-borne sound transmission around or beyond the barrier via rigid, solid connections to the barrier—particularly at the edges.

Rarely will a barrier perform as well in an actual field installation as in a laboratory test, since laboratory test procedures carefully eliminate all significant *unintentional* paths. Even a tiny hole, opening, or shrinkage crack may seriously degrade the performance of a partition or panel. Significant energy transmission to the connecting structure may occur in some types of constructions.

Surprising as it may seem, *one square inch* (6.5 cm²) of hole through a 40-dB wall will transmit as much acoustical energy as almost 100 *square feet* (10 m²) of the wall! Thus, a shrinkage crack 0.01 in (0.25 mm) wide and 12 ft (4m) long can reduce the Sound Transmission Loss of a 40-dB wall, 8 ft × 12 ft (2.5 m × 4 m) in area, by as much as 3 dB; the same crack can degrade a 50-dB wall of the same area by almost 10 dB! The better the wall, the more serious the leak. (See Figure 17.)

A multitude of tiny leaks can also add up to a serious problem. The STC-rating of a lightweight, porous concrete block wall, for example, often can be improved by 5 to 8 points simply by sealing both surfaces tightly with paint.

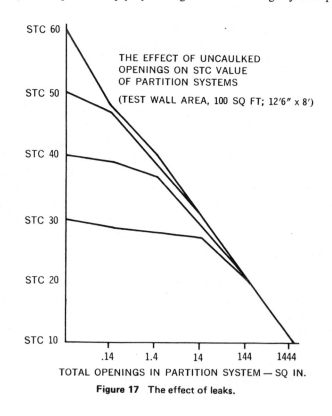

Figure 17 The effect of leaks.

The perimeter of a typical 36 in × 80 in (1 m × 2.25 m) door, fitted with only $\frac{1}{8}$ in (3 mm) clearance on all four edges, may provide up to 30 in^2 of opening unless the door is thoroughly gasketed on all four edges. The typical undercut for return air renders a door almost useless as a barrier; and a relief grille or louver is even more disastrous. The inevitable perimeter "leak" around pipes, conduits, and ducts which penetrate walls is also a major problem unless sealed.

Remember, if air can get through, sound will!

A standard $1\frac{3}{4}$ in. (4.4 cm) hollow-core wood door, even if well-fitted and tightly gasketed, provides an STC-rating of only STC-26. Therefore, it represents a "weak link" in almost any good wall.

A good "rule of thumb" for designing a "balanced" construction (wall plus any tight, well-caulked, or gasketed fenestration) is:

1. If the area of the door (or window or other low Transmission Loss insertion) is *less than 25%* of the total wall area in which it is inserted, its STC-rating may be up to 5 points lower than the STC of the wall without seriously degrading the barrier's performance.
2. If more than 25% and *less than 50%*, its rating may be 2 points lower than that of the wall.
3. If *more than 50%*, its STC-rating will essentially determine the performance of the entire barrier.

Windows, therefore, represent a major leak in almost any reasonably good wall. Special glass and glazing techniques are required to prevent serious sound transmission through large glazed areas.

As discussed earlier, a panel usually vibrates in several modes, with a significant shear wave (like the shaking of a rope) carrying considerable acoustic energy to the edges of the panel. Thus, the attached surfaces (walls, floors, etc.) may be set into vibration, radiating their energy into adjacent and distant areas. In other words, the sound can "flank" the barrier.

Figure 18 shows how the shear wave in a barrier can set attached surfaces into motion similar to its own, causing them to become sound sources, too.

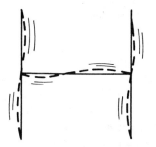

Figure 18 Transmission by flanking.

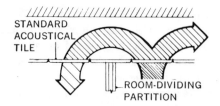

Figure 19 Flanking through acoustical ceilings.

Lightweight, continuous curtain walls, thin diaphragm floors (thin concrete on bar joists), balloon-framed wood construction, and similar systems are particularly susceptible to this defect. Carefully planned discontinuities, resilient connections at perimeters, and other design devices to interrupt these transmission paths are usually successful in controlling such problems. (See Section III, page 124), for suggested construction details to minimize "leaks and flanking.")

One very common and potentially serious leak and flanking path in many buildings is the lightweight, mechanically suspended acoustical ceiling (particularly the "lay-in" panel type, with continuous runners). Often the partitions are erected to (not through) the ceiling. (See Figure 19.)

The light, porous tiles and panels are relatively poor barriers at best; and a potentially serious leak often occurs at the joint between the ceiling and the top of the partition. If the metal supporting runners are continuous over the top of the partition, a shear wave, capable of transmitting energy to adjacent spaces, often develops in the ceiling diaphragm. Such construction may almost vitiate even the best wall or partition between spaces.

The "over-the-top" path must provide at least as good a barrier as the wall or partition if a balanced construction is to be achieved. (See Section III, page 128, for details and suggestions to minimize this problem.)

Good acoustics result from infinite attention to details—even apparently trivial details. Nowhere is this more important than in dealing with leaks and flanking. Even very good constructions can literally be ruined by the traditional minor carelessnesses and overlooked details all too typical of the building industry. While a wall construction, for example, may unquestionably be capable of providing STC-50, a careless draftsman or sloppy mechanic can turn it into an STC-25 failure.

Partial-height Barriers Sometimes total enclosure is impossible, and only screens or partial-height barriers are feasible. Still, considerable shielding can be provided by a barrier which is sufficiently large, heavy, and tight. A solid panel, weighing not less than 2 lb/ft^2 (10 kg/m^2), without holes or openings, can be reasonably satisfactory. Partial-height office dividers, "Sunday School screens," and similar barriers have been used for years with some degree of satisfaction.

For practical purposes, shielding provided is less than 20 dB in the higher frequencies, and diminishes to less than 5 dB in the lower frequencies. Yet, this may be the margin between complaints and satisfaction in many cases.

To be useful, the barrier must be wider than its height, and the height must be not less than 66 in (1.7 m), with the "receiver" (being shielded) located near the barrier, or with the "source" (being shielded against) located near the barrier. Line of sight between source and receiver must be cut off completely if any shielding is to be expected.

Occasionally, high (over 20 ft [6 m]), long walls are erected to protect adjacent homes against plant noise; and heavy, high partial enclosures are erected to reduce noise from large power transformers, etc. Obviously, such barriers are expensive and are not simple solutions; their design should be undertaken with care and with the assistance of qualified consultants.

Open screens, low picket or slat fences, and rows of shrubbery are almost useless—except for appearance.

Sound Absorption   Sound absorbents are used:

1. To reduce the Sound Pressure Level within spaces.
2. To prevent reflections from surfaces.
3. To control reverberation within spaces.

In general, absorbents should be located:

1. As near the offending sound source as possible.
2. On surfaces producing unwanted reflections.
3. On surfaces not required for helpful reflections.

In practice, this usually means that they are found on ceilings, walls, and floors of rooms; on panels surrounding noisy equipment; within the cavities between wall or partition surfaces, and the like.

(See Section III, page 133, for criteria and recommendations for amount and location of absorbents within spaces.)

Caution: It is very important to recognize that most acoustical absorbents are very poor sound barriers. They are usually porous and lightweight—quite the reverse of what is required to reflect or isolate sound. Therefore, it is usually quite useless to apply acoustical tile, for example, over a wall to reduce sound transmission through the wall. The added mass is trivial, and the innumerable "holes" through the absorbent provide good paths for sound transmission. Only the energy which is absorbed is eliminated, and it will usually be less than 5 dB. When, for example, isolation of 30 to 60 dB is required, absorbents are not very useful.

The Mechanism of Sound Absorption   Absorptive materials are "transducers"—that is, they convert acoustic energy to a different form of energy, usually heat. The mechanism of conversion is different for each of the different types of absorbers, but the result is the same—some acoustical energy is "lost" when the sound wave reaches the absorber. A description of the structure and the energy conversion process for each of the three major types of absorbers is given later in this section.

Sound Absorption Coefficient and Noise Reduction Coefficient The ratio of acoustic energy absorbed (converted to heat) to the energy incident upon a surface is called the "Sound Absorption Coefficient" of the material.

If one square foot of open window is assumed to transmit all and reflect none of the acoustical energy which reaches it, it is assumed to be 100% absorbent. This unit—one square foot of totally absorbent surface—is called a "sabin." Then the absorption of one square foot of an absorptive material is compared with this standard, and the performance is expressed in coefficients, such as 65% or .65. As will be discussed later, this rating method is not as simple or logical as it may appear, but workable measurement procedures have been devised to compare and rate absorbents and to predict their performance in actual use.

There is no practical way to measure directly the sound absorption of materials with randomly incident sound (as normally would occur in a building) or on special mountings (such as mechanical mounting systems with a deep air space behind the surface material). The most common and useful test method is the "Reverberation Room Method"—ASTM Method C 423-66. A large panel of the material is installed in the test chamber in somewhat the same manner as it would be installed on the job. Then the Absorption Coefficients (often called the "Sabine Coefficients") are computed from the effect of the test specimen on sound decay in the room, using the Sabine reverberation time formula (see Section II, page 57). The results of this test method probably relate reasonably well to the performance of the materials in actual job installations. Sound Absorption Coefficients are usually determined at 125, 250, 500, 1000, 2000, and 4000 cps (Hz).

In the United States, a well-established practice is to average (arithmetically) the Sound Absorption Coefficients measured at 250, 500, 1000, and 2000 cps (Hz), and to label this figure the *"Noise Reduction Coefficient"* (NRC) for the material tested. It has no physical meaning and should not be used indiscriminately; but, like many one-number rating systems, it has some practical validity. It is a reasonably useful means of comparing *similar* materials and of predicting the effect of the material in reducing general, broadband noise within ordinary rooms, offices, and the like.

(For a discussion of Reverberation Time and Noise Reduction calculations, see "Design Formulas," Section II, page 57.)

Porous Absorptive Materials The best-known acoustical materials are the absorptive materials. Porous, "fuzzy," fibrous materials; perforated boards; and similar building products are widely known as "acoustical." The normal furnishings in a room are also highly absorbent. For example, fabrics, carpets, cushions, and upholstery may be very effective absorbers.

Porous absorptive materials can be described as a matrix of fibers, granules, or particles of some sort. The air contained within the matrix is "pumped" back and forth within a restricted space when sound energy reaches the material. As the air moves within the matrix, frictional losses occur as heat, and the acoustic energy is reduced accordingly.

Figure 20 shows (enormously magnified) the internal structure of a typical porous acoustical absorbent material.

It is imperative that the internal structure be composed of interconnected pores and voids. Only open-cell structures are effective absorbers. Many plastic and elastomeric foams and most glass and ceramic foams tend to have closed, nonconnected voids. Air movement within them is very limited—almost nonexistent in most cases—and they provide little or no sound absorption. A simple test of a material is to blow smoke through it. Those materials which pass no

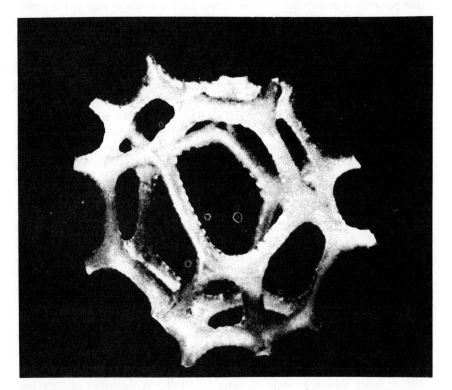

Figure 20 Photomicrograph of open-cell foamed urethane. (Courtesy Foam Division of Scott Paper Company.)

smoke will provide almost no absorption. If they pass smoke too readily, with little pressure, they may not have good absorption either.

The amount of energy conversion is determined by the resistance to airflow within the matrix (more correctly, to the "impedance," since the flow is AC rather than DC, and is rarely in phase with the pressure causing the flow; however, the principle remains the same). As might be expected, if the resistance is too low, frictional losses are low, and little energy conversion occurs. If resistance becomes too high, flow becomes so restricted and air motion is so limited that frictional losses are low. However, there is a relatively broad range for acceptable flow resistance, and optimum resistance is not sharply critical in most materials.

It is important to remember that the maximum excursion of each molecule of air is usually relatively small; it is not necessary that the air enter or leave the absorptive medium—only that it move within the medium. Also, air movement can take place in any direction—vertically, horizontally, diagonally—within the matrix in which it is confined.

The porosity of absorptive media is usually related to the density of the media. However, this relationship is unique to each material or type of material. Density alone is not a meaningful indicator of the absorptive effectiveness of even a family of similar materials.

There is a predictable relationship between sound frequency and absorption, since motion of the individual molecules (reversals of direction of each molecule) is directly related to the frequency. Thus, the dimensions of the material compared with the wavelength of the sound reaching it are significant. Theoretically, maximum absorption occurs when absorber thickness is about one-quarter the wavelength of the lowest frequencies to be absorbed. Therefore, thick materials are required to absorb low-frequency sound, while even thin materials will absorb high-frequency sound. In practice, this usually means boards, panels, or blankets ranging in thickness from $\frac{1}{2}$ in (1.3 cm) to about 4 in (10 cm).

It should be obvious that mere rough textures, such as sand float finish on plaster or some of the so-called "acoustical paints" which repeatedly appear on the market to deceive the gullible, cannot be effective acoustical absorbers. The wavelength of even the very high frequencies of interest in most work is many times greater than the maximum dimension of such irregularities. But even more important, such products are not porous, and there is no opportunity for any airflow within them. Therefore, there is no mechanism for sound absorption, and these products are almost totally useless.

Mere rough textures—even fieldstone or patterned brick—do *not* "break up" the sound; sound is not fragile or brittle, and it is merely reflected from hard surfaces, even fairly deeply textured surfaces.

The designer should be equally skeptical of the performance of very thin porous or textured materials. A thin flocked surface over hard and impermeable materials is almost useless except at very high frequencies. Very thin layers of porous foams have frequently been sold as sound absorbers. Some such materials are actually very good absorbers in thicknesses of one inch (2.54 cm) or more; but in the $\frac{1}{8}$ to $\frac{1}{4}$ in. (3 to 6 mm) thicknesses in which they appeared on the market, they were quite poor absorbers in any frequencies below about 1000 cps (Hz).

In practice, the significant characteristics of acoustical absorbents include far more than just the acoustical absorption. Cost is the most significant parameter; strength, hardness, durability, cleanliness, weight, maintainability, fire resistance, moisture resistance, and appearance are among the other characteristics for which the designer must look. Hence, some of the best absorptive materials are not well-suited for actual use unless modified in various ways.

One of the most effective modifications is to provide a surface of some sort and to depend upon the surface to protect the absorptive material and to receive the maintenance efforts, whether such efforts be painting, washing, or other normal maintenance procedures. Many acoustical absorbents, when used in exposed locations, are protected with porous or perforated facings of various types. The facings, then, introduce another significant factor. Any facing of any type will affect the acoustical performance of the material in some way. Usually the facings tend to degrade the high-frequency performance of the absorbent material (above 1000 cps [Hz]) but they often improve the low-frequency performance. An examination of Figure 21 will explain this phenomenon.

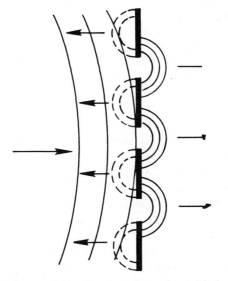

Figure 21   Diffraction through a perforated facing.

When the solid or reflective areas between perforations become large enough, these areas reflect much of the sound which strikes them, particularly the higher frequency energy. However, the lower frequency sound, with its longer wavelength, can diffract around the reflective areas and provide the necessary pressure behind the surface to activate the air enclosed in the absorptive material behind the surface. In effect, a plate with a number of controlled orifices is inserted between the oncoming wave front and the absorptive material.

The resistance of a perforated facing over an absorptive material must vary with the thickness and resistance of the absorptive material behind it to give maximum absorption over the widest range of frequencies. However, as the per cent open area decreases and hole spacing and size increase, high-frequency absorption decreases, low-frequency absorption tends to increase, and an increasingly sharp absorption peak appears.

Perforated facings used over commercial acoustical absorbents vary from about 5 to 40% open area, depending upon the thickness of the facing material, hole size, and hole spacing. Even very porous fabrics (such as speaker grille cloth) are used, and ordinary #16 fly-screen is frequently employed for this purpose.

A common type of acoustical absorptive tile used in architectural work consists of a low-density fibrous insulation board with a thick, heavy, painted surface. Then holes are drilled through the surface into the absorptive material below. Airflow takes place in all directions, but maximum airflow appears to be horizontally between the cylindrical voids in the structure.

Even a solid, unbroken film can be used over a good absorbent matrix if the film is thin enough, light enough, and flexible enough to impose little resistance between the impinging sound wave and the air in the matrix. Very successful acoustical tiles and panels are available with flexible films of Mylar, plasticized vinyls, etc., ranging in thickness from $\frac{1}{2}$ to 2 mils. The films are *not* attached solidly to the entire surface they cover, since this would cause them to form a very rigid covering over the tiny openings in the surface of the material below. Rather, they are attached at the edges of the tiles or panels or in a few spots or in widely spaced strips so that the entire film is free to flex and to be relatively "limp." As mentioned earlier, the air at the surface of the panel need not enter the matrix, nor need the contained air flow out of the matrix. It is only necessary that the pressure of the oncoming wave be imposed on the contained air in the matrix to cause it to move. A thin, flexible, limp layer over the surface will permit this pressure transfer.

Since painting is a common, standard maintenance method, it is important that its effect on the absorption of the acoustical absorber be known. As explained above, a rigid, intimately attached film will cause the material to reflect rather than absorb sound. For this reason, many acoustical tiles or panels are made with surfaces containing large openings—holes, fissures, etc.—to permit painting without degrading the sound absorption of the material. In practice, such surface openings should provide about 15% to 18% open area, and they should be large enough to prevent "bridging" or filling by normal paint used in normal application techniques.

The method of attaching or supporting absorbents has an appreciable effect upon their performance. (See Figure 22.) Probably most materials are adhered directly to a hard, impervious surface. In architectural practice, however, panels or tiles are often suspended on runners or furring strips, with an air space behind them. The effect of the air space is to increase the low-frequency absorption considerably and to degrade slightly the higher frequency absorption. The absorption curve tends to rotate about the value at about 500 cps (Hz).

For this reason, published values of acoustical absorption of commercial materials are always related carefully to the method of application.

Depending upon the particular material being considered, the optimum air space behind the panel varies from about 2 in. (5 cm) to over 12 in. (30 cm). However, little change takes place after the air space reaches 16 in. (40 cm). As a result, the principal commercial testing laboratory for acoustical absorbents in the United States has standardized on a 16 in. (40 cm) air space for all tests on the No. 7 Mounting—the so-called mechanical suspension mounting system.

There are literally hundreds of absorbent materials in use today. They range from hair felt to upholstered seats. (For a complete up-to-date listing of materials, refer to the current annual *Sweet's Catalog* or similar publications.)

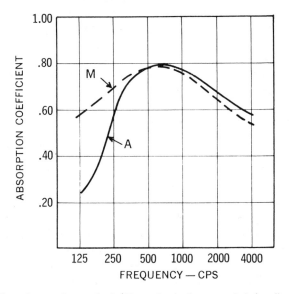

Figure 22 Effect of mounting method. (M, mechanically suspended; A, adhesively applied.)

A brief summary of the principal types of materials follows:

TABLE 7 Common Porous Absorbents

Material	Thickness (in)	(cm)	Density (lb/ft^3)	(kg/m^3)	Noise Reduction Coefficient
Mineral or glass wool blankets	$\frac{1}{2}$ to 4	1.3 to 10	$\frac{1}{2}$ to 6	8 to 100	.45 to .95
Molder or felted tiles, panels, and boards	$\frac{1}{2}$ to $1\frac{1}{8}$	1.3 to 3	8 to 25	125 to 400	.45 to .90
Plasters	$\frac{3}{8}$ to $\frac{3}{4}$	1 to 2	20 to 30	320 to 480	.25 to .40
Sprayed-on fibers and binders	$\frac{3}{8}$ to $1\frac{1}{8}$	1 to 3	15 to 30	240 to 480	.25 to .75
Foamed, open-cell plastics, elastomers, etc.	$\frac{1}{2}$ to 2	1.3 to 5	1 to 3	16 to 48	.35 to .90
Carpets	Varies with weave, texture, backing, pad, etc.				.30 to .60
Draperies	Varies with weave, texture, weight, fullness, etc.				.10 to .60

In particularly difficult environments, the more conventional absorbents are usually unsuitable. Very high temperatures, corrosive fumes, high humidity, dust, abrasion, impact and other physical abuse, and even very high Sound Pressure Levels (above 150 dB) often eliminate the more common materials from consideration. Fortunately, however, there are available materials or combinations of materials which can be used under such circumstances.

Metal "wools" made of stainless steel or copper fibers have been used where high temperatures or corrosive gases are encountered. Unfortunately, they are very expensive, and they must be used in thick layers to be very effective.

Porous ceramic materials and some plastic-bonded mineral or glass wool fibers withstand humid conditions reasonably well. Foamed, open-cell urethanes are particularly useful in such environments.

Very few absorbents can withstand much abrasion or impact. A porous, sturdy facing is usually required to protect the absorbent in such applications.

Surprisingly, even many resilient materials with high-temperature resistance do not perform well under sustained exposure to very high Sound Pressure Levels. At about 160 dB (roughly, the levels at the exhaust of a jet engine), binders burn out of most bonded mineral or glass wools; and the brittle, fragile structure of most boards or panels is subject to cracking or disintegration under the flexing and vibration which occurs within the material at these levels. Open-cell (reticulated) urethane foams perform very well under these circumstances, and they are frequently used in such applications (where they are not exposed to hot exhaust gases or other high-temperature environments).

Massive, strong ceramic or porous masonry block materials are often used where levels do not exceed 150 dB, but their absorption coefficients are relatively low when their structure is dense enough to resist the forces involved.

Porous, sintered metal is occasionally used for special applications, but it is extremely expensive, and it is rarely used in large quantities.

In special rooms, such as anechoic test chambers, tapered "wedges" up to 60 in. (1.5 m) length are used. Their design is a highly specialized procedure and should be undertaken with great care.

Occasionally, so-called "space absorbers" have been tried where the more usual panels, tiles, or blankets are not practicable. These free-hanging units may be thick panels (one type was 24 in. X 48 in. X 2 in. [0.6 m X 1.2 m X 5 cm]) of any of several types of absorptive material: hollow tetrahedrons formed of dense glass wool blankets covered with a thin vinyl film; two molded hollow wood fiber cones attached at their base to enclose a large volume of air; hollow cylinders of glass wool with a perforated metal facing; etc. The units usually exhibit a very high absorption per unit surface area, and they can be hung where required. However, they are relatively expensive, and they have not been used widely.

Most upholstery, seat cushions, fabrics, and clothing are absorptive. Published data are often available from manufacturers (particularly manufacturers of auditorium seats). People—the audience present—in a room are highly absorbent. Their absorption is always taken into account in design of critical spaces.

Diaphragmatic Absorbers When a wave front impinges upon a panel, the panel vibrates at the same frequency as the sound reaching it (or at some harmonic). Since the panel material is never perfectly elastic, some energy is lost because of the inherent damping in the panel or the assembly in which the panel is used. This energy loss can be usefully employed as sound absorption in many cases.

As might be expected, low-frequency sound will move panels more effectively than high-frequency sound, since the impedance match is usually much better between the air and the panel. High-frequency sound tends to be reflected without losing much of its energy to the panels normally used for most purposes.

In practice, thin sheets of metal, plywood, plastic, or even paper have been used as diaphragmatic absorbers. (See Figure 23.) Typical units include vacuum-formed ceiling panels of thin styrene or vinyl, damped sheet metal, and even plywood. Because the absorption is significant principally in the lower frequencies, such absorbers are normally used only to supplement other absorption or to absorb specific low-frequency sounds. This is particularly useful in many applications, since it is often impractical to use the extremely thick layers of fibrous absorption required for good low-frequency absorption, but it is simple to use combinations of panels and blankets to provide good broadband absorption.

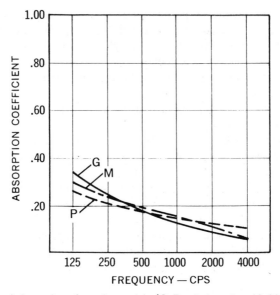

Figure 23 Sound absorption of panel materials. (G, Ds window glass; M, 26 ga. sheet metal; P, 3/8″ plywood.)

Thin sheets of plywood over a confined air volume can also provide useful low-frequency absorption in building construction. If porous mineral or glass wool blankets are hung within the cavity behind the plywood surface, the absorption is appreciably increased. A design procedure for use of this type of construction has been worked out, and it can be quite useful for special applications such as theaters, music rooms, radio studios, and the like.

One remarkably efficient diaphragmatic unit commercially available today is a molded panel consisting of glass fiber, bonded with a plastic material to make a shell of about $\frac{1}{8}$ in. (3.2 mm) thickness. The individual units are shaped into shallow, pyramidal vaults about 24 in. × 24 in. (0.6 m × 0.6 m), and erected on metal runners to provide a substantial air space behind them. They provide a remarkably flat absorption curve, with unusually good coefficients from 125 cps (Hz) to 4000 cps (Hz).

Absorption varies with the mass and stiffness of the panels. Hence, it is difficult to calculate or forecast the absorption of panels unless all design details and dimensions of the completed construction are known accurately. In practice, prototypes are tested in a reverberation chamber to determine actual performance.

In general, practical requirements of strength, damage resistance, cost, and other characteristics of building and equipment materials tend to limit the applications for diaphragmatic absorbers. However, it is often wise to consider the inherent absorption of any such thin panels during design, both as a means of supplementing other absorption and as a major consideration in spaces where low-frequency absorption may even be undesirable (music spaces, organ lofts, etc.).

Resonant Absorbers   Resonators (often called Helmholtz resonators) are cavities which confine a volume of air which communicates with the atmosphere by means of a small hole or channel in the surface of the cavity. If the dimensions of the cavity are very small compared with the wavelength of sound reaching the opening of the cavity, the resonator "tunes" to a specific frequency. The fundamental vibration of the confined air volume is a periodical airflow through the channel into and out of the cavity, and the air in the cavity acts as a spring. The kinetic energy of the vibration is essentially that of the air in the channel moving as an incompressible and frictionless fluid.

In practice, this type of absorber has limited application, since its peak absorption is a narrow band of the lower frequencies of interest in most sound control work. However, for applications where it is important to get high absorption of low frequencies, resonators can often usefully supplement other absorbents.

One of the most successful types of resonators used in the building industry is the ordinary concrete block (see Figure 24) with carefully designed slots cut into one face to form the channel which communicates with the hollow cells within the block. Most concrete masonry block used in building today is rather porous and somewhat absorbent. As a result, the blocks used as resonators have a distinct absorptive peak, usually in the frequencies between 100 cps (Hz) and 300 cps (Hz), with some useful absorption in the frequencies above 300 cps (Hz). (See Figure 25.)

Figure 24   Concrete block resonator. (Courtesy The Proudfoot Company, Inc.)

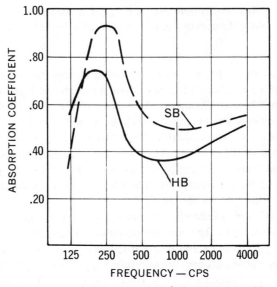

Figure 25 Absorption of concrete block resonator. (HB, hollow cells; SB, cells stuffed with wool.)

If an absorbent material, such as mineral or glass wool, is inserted into the cavities, the absorption peak is effectively broadened, and the absorption in the higher frequencies is increased significantly.

Occasionally, resonators are attached to stacks, ducts, pipes, or other structures in which a strong, low-frequency tone must be attenuated. In such applications, it is often possible to obtain considerably more noise reduction with a resonator than with very thick layers of fibrous absorbents. The principle is similar to so-called side-branch resonators used for exhaust mufflers and other applications where nondissipative absorption is desirable. Design of such resonators is governed by the formulas for resonance of Helmholtz resonators which can be found in almost any physics textbook. The frequency (or center frequency of the band of frequencies) to be attenuated is determined, and the volume and dimensions of the resonator and the orifice and neck or channel into the resonator are chosen to produce peak absorption at that frequency.

In most building applications, however, the dimensions of the cavity and the channel are governed by the dimensions of the wall structure into which they are built or by practical limits to the dimensions of the building units used. The design parameters which can be manipulated are normally the length and width of the slot or slots into the cavities.

It is usually imperative to choose practical units or materials with which to form the resonators, design proper openings of channels, and then test the units in a reverberation chamber to compare their performance with the calculated performance. Modifying the measured performance tends to be an empirical process because of variables inherent in the materials, shapes, and other characteristics of the units.

Design Formulas It is possible to calculate with reasonable accuracy the effect of introducing absorption into a space. There are two formulas of particular interest to the engineer and architect.

Noise Reduction

$$NR = 10 \log_{10} \frac{A_o + A_a}{A_o}$$

where

$\qquad A_o$ = original absorption present in sabins;
$\qquad A_a$ = added absorption in sabins;
$\qquad NR$ = Sound Pressure Level reduction in decibels.

(**Note:** The surface area in square feet multiplied by the absorption coefficient = the sabins of absorption. When added absorption covers an existing surface, the coefficient of the added absorption must be reduced by the coefficient of the existing surface covered by the added absorption. Here we have the familiar $10 \log_{10}$ times a ratio. The ratio of absorptive area "before-and-after" is obviously related to the ratio of the amount of energy absorbed.)

It is obvious that in a highly absorptive space, the effect of the added absorption is small compared with its effect in a nonabsorptive space. In practice, less than 10 dB noise reduction can be accomplished in most installations by the introduction of absorption alone.

Reverberation Time

$$T = 0.05 \frac{V}{A}$$

where

$\qquad T$ = time in seconds for the Sound Pressure Level to decay 60 dB
$\qquad\qquad$ after the source ceases;
$\qquad V$ = room volume in cubic feet;
$\qquad A$ = total absorption in sabins within the space.

or:

$$T = 0.161 \frac{V}{A}$$

where

T = time in seconds for the Sound Pressure Level to decay 60 dB
after the source ceases;

V = room volume in cubic meters;

A = total absorption in metric sabins within the space.

This is the well-known Sabine formula, developed somewhat empirically by Professor Sabine more than a half-century ago. While much discussion and many modifications to the formula have filled pages of the literature ever since, the formula continues to be as accurate and useful as measurement and calculation techniques can justify. Since reverberation room measurements of absorbents employ this formula to compute the coefficients of the test specimens, it is highly likely that the formula is better suited to design calculations than any other formula.

Room reverberation is actually somewhat more complex than the simple formula would suggest. The formula assumes that the wave front moves fast enough to reach many room surfaces quickly, losing some energy at each reflection. (Actually, the formula assumes that all surfaces will be reached by each sound pulse eventually.) Many studies of many rooms have shown that the reverberant pattern within spaces is highly complex and not actually completely predictable. Nevertheless, the formula is a good guide tool for the experienced practitioner—and a possible snare and delusion for the inexperienced.

It is important to remember that the absorbent materials do not "reach out and grab" nor do they "suck up" the sound energy. They can absorb only the energy which reaches them. The situation is somewhat analogous to light. If a 60-W light bulb were burning in a glistening white bathroom, the room would be very bright because of the reflective walls. If one began painting walls black, one by one, the room would become increasingly dark, although the same light source would be emitting the same amount of energy. It should be obvious that a small patch of absorption in the vicinity of a major sound source is almost useless in reducing the sound level within a space; yet this mistake is observed constantly.

Vibration Isolation and Damping Since all materials are somewhat elastic, all materials transmit sound and vibration to some degree. Many materials used by engineers and designers have a high degree of elasticity; hence, such materials may transmit sound vibration readily. Often this characteristic of materials is a serious nuisance. Minimizing such transmission often presents a greater challenge than any other aspect of noise control.

Even the most elastic material is not perfectly elastic; some energy is lost as the material oscillates. This energy loss can be used to control vibration, just as the elasticity of the material can be used to isolate or "float" a vibrating source. An explanation of this apparently contradictory situation follows.

The Mechanism of Isolation and Damping An obvious control technique would be to interrupt the transmission path; and this is often a practical and successful approach. However, knowing how to accomplish this appears not to be so obvious. Many engineers and designers appear to misunderstand the mechanism of vibration isolation or to harbor misconceptions about the concept.

Generally, two problems are involved in vibration control:

1. Preventing energy transmission between the source and the surfaces which radiate the sound and vibration.
2. Dissipating or attenuating the energy somewhere in the structure being considered.

The first problem involves assembling the structure in such a way that energy transfer through connections is inefficient. The term often used to describe this process is *"impedance mismatch."* Unfortunately, "impedance" seems to convey little information to many engineers. Perhaps an examination of this term will clarify the matter and help to explain how this concept can be used in design.

The acoustical impedance of a material is found by multiplying its density by the velocity of sound in the medium; this results in impedance units such as lb/in.^{2}-sec or in.-lb/in.^{3}-sec (kg/m^2-sec). The first expression represents force per unit area per second; the second represents energy per unit volume per second. Both indicate a rate at which force can be applied per unit area or energy can be transferred per unit volume of material. In other words, some materials cannot accept energy as fast as others. A good, homely analogy would be trying to hammer air or stone; it is simply impossible to hit the air as hard as the stone with a hammer. It is pretty difficult to make much noise hammering the air, but it is easy to make a lot of racket pounding a stone.

The second approach involves accepting the energy into the system and converting it to heat before it can be radiated to the surroundings; this is called *damping*.

TABLE 8 Acoustical Impedance of Various Materials

Material	Acoustical Impedance	
	(lb/in.^{2}-sec)	*(kg/m^2-sec)*
Cork	165	116 X 10^3
Pine	1,900	1,340 X 10^3
Water	2,000	1,410 X 10^3
Concrete	14,000	9,870 X 10^3
Glass	20,000	13,400 X 10^3
Lead	20,500	14,450 X 10^3
Cast Iron	39,000	27,500 X 10^3
Copper	45,000	31,725 X 10^3
Steel	58,500	41,240 X 10^3

In vibration control, low-impedance materials are inserted between high-impedance materials to interrupt the transmission paths. Even if the interrupting materials are highly elastic, they cannot transfer the energy of the oscillating or vibrating source fast enough to transmit much noise or vibration. Thus, a steel spring can support a vibrating machine on a concrete base without transmitting much energy between them. The spring simply stores most of the energy which it accepts, transmits a little, dissipates a little, and returns most to the vibrating system with each cycle.

If a mass, supported on a perfectly elastic spring resting on an infinitely stiff and massive support, were set into vibration, it would oscillate at a rate determined only by gravity and the *spring rate* (load to produce unit deflection) of the spring. The frequency of this oscillation (*"natural frequency"*) will be:

$$f = 3.13 \sqrt{\frac{1}{d}}$$

where

f = frequency in cps (Hz);
d = static deflection of the spring in inches under the load imposed
    (determined by the stiffness of the spring).

or:

$$f = 5 \sqrt{\frac{1}{d}}$$

where:

f = frequency in cps (Hz);
d = static deflection of the spring in centimeters under the load imposed
    (determined by the stiffness of the spring).

If the spring were truly "perfectly elastic," the mass would vibrate indefinitely at the "natural frequency" of the system (if the base were truly infinitely stiff and massive). The spring would simply store the kinetic energy of the system as potential energy and return it to the system with each cycle. However, no components of this nature exist. Internal losses within the system will eventually cause the free vibrations to cease unless outside force is applied to the mass. The decay of the vibration will, of course, be roughly logarithmic (see Figure 26), since each oscillation will be damped by some ratio of the energy or amplitude of the previous oscillation.

The energy lost in each cycle in internal dissipation within the spring is called "damping." (Of course, some energy is transmitted to the support; this, too, is ultimately dissipated and lost to the system.)

Damping is a means of dissipating or attenuating vibrational energy. In a sense,

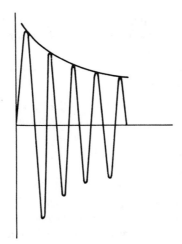

Figure 26 Decay of a damped oscillation.

the damping mechanism is the opposite of resilient isolation. Instead of storing energy and returning it to the vibrating system, damping materials accept the energy and convert it to heat. The two mechanisms should not be confused.

Most vibrating systems can be thought of as a mass supported on a spring with an attached dashpot or other damping device. (See Figure 27.)

The damping may be:

1. Viscous—with force directly proportional to velocity, such as a dashpot with fluid forced through an orifice.
2. Hysteresis—"structural" damping which depends upon displacement with little frequency effect.
3. Coulomb—frictional damping.

Nearly every text on dynamics contains a detailed treatment of the subject of vibration and damping. Vibration isolation and control are covered at length in

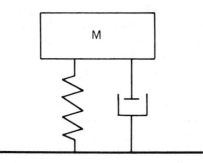

Figure 27 A simple vibrating system with damping.

the References and Bibliography (Section III). However, the engineer and designer need know only a few basic concepts to obtain a reasonable understanding of damping and damping materials.

For several reasons, damping is a fortunate characteristic of all materials. The two most important reasons are:

1. If force is continually applied to a resiliently supported mass, the system will continue to vibrate at a given rate and amplitude. Should the system be vibrating at its natural frequency, it will "resonate"—that is, the amplitude of vibration will increase with each cycle. In a perfectly elastic system, this increase would approach infinity.

Actually, very dangerous resonance conditions can and do occur in practice. Fortunately, the inherent damping in any system will eventually limit the motion of the system to some maximum. For resiliently supported machines, there is usually a brief period of resonance as rotational speed increases from zero to operational rates. Either positive restraints must be provided or internal damping of the system must limit the motion. Of course, the greater the damping, the less the isolation. Too much damping degrades the performance of the mounts, while too little may permit undesirable movement in the system.

So-called "shock absorbers" on an automobile are typical of practical "dampers." They limit motion of the sprung masses, and they restore the system to stability.

(**Caution:** As can be seen from Figure 28, serious resonance can occur when the frequency of the supported vibrating equipment ("forcing" frequency) coincides with the "natural" frequency of the resilient mounting system. Isolation occurs only when the ratio of forcing frequency to natural frequency exceeds 1.4/1. In practice, this ratio is usually set at 3/1 or over. See particularly Section III, page 164, for design procedure for resilient mounting systems.)

2. If a panel of elastic material—metal, wood, glass, etc.—is set into vibration by almost any source, it will radiate sound. If such a panel is part of the housing for a rotating or vibrating machine, for example, the panel can act as an efficient "loudspeaker," amplifying the sound of the machine many times. Or a panel, if struck, will "ring" for some time. Fortunately, these vibrations can be restricted and damped by proper design of the panel and by the application of damping materials which dissipate the energy as heat.

Resilient Mounting Materials   Properly designed steel springs continue to be one of the best "resilient materials." However, high-frequency vibrations can travel through the spring, even while low-frequency vibrations are well-isolated. Therefore springs are usually used in conjunction with elastomers and similar materials.

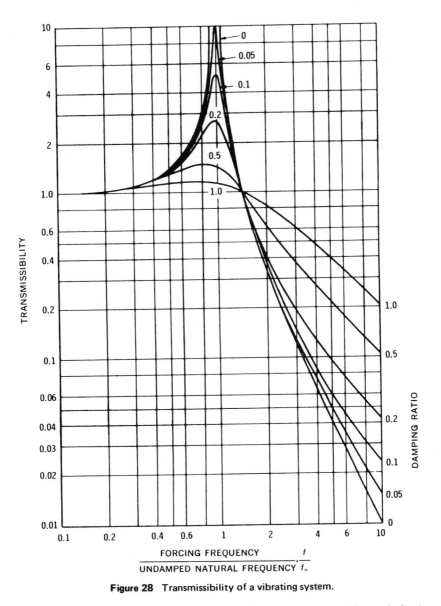

Figure 28 Transmissibility of a vibrating system.

Cork, hair felt, bonded glass fiber boards, solid elastomers, and foamed plastics and elastomers are often used for resilient mounting systems. Usually, they are used in compression, either as pads or blocks of various sizes and thicknesses. The elastomers are often used in shear, since their performance is usually more truly elastic in this mode.

Internal damping is very high in many of these materials, particularly the granular and fibrous products. As a result, rarely can they provide much vibration isolation for frequencies below about 10 to 12 cps (Hz).

Most such materials exhibit strongly nonlinear stress–strain characteristics. Usually they are "hardening" springs, with the spring rate increasing with the deflection. Occasionally, they are "softening" (or even collapsing) springs, with spring rate decreasing or reaching a constant value. The dynamic performance of such materials is usually far different from their static performance; usually the spring rate is much higher under dynamic loading. Figure 29 shows a typical curve.

The large area within the "hysteresis loop" represents the energy loss per cycle; this, of course, is the damping provided.

Most such materials may be loaded from 1 lb/in.^{2} (7 kPa) to over 50 lb/in.^{2} (350 kPa). If loaded too heavily, they tend to fatigue and to experience permanent "set" or deformation. The useful life for these products appears to be about 20 years, although cork and some natural rubbers tend to harden and lose resiliency in much less time.

Loadings on these materials must be reduced, often to one half, when they are subjected to shock.

For very high unit loadings and continuous shock loads, special materials are available. For example, multiple layers of fabric impregnated with Neoprene are laminated into sheets or pads. Such products may be loaded to 500 lb/in.^{2} (3500 kPa) or more.

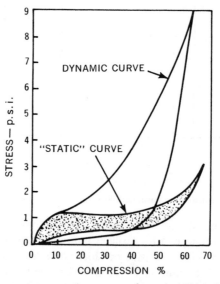

Figure 29 Dynamic response vs static response of a typical "high-damping" elastomer.

Damping Materials Materials used solely to damp vibrations are usually visco-elastic substances, ranging from filled bitumens to specially formulated elastomers. Some of the new polymer plastics are particularly effective.

Materials are available in sheet form for adhesive application or for vulcanizing to metal; as liquids or thick fluids for spray or trowel application; and as tapes, often with contact adhesive already applied.

Flexing of the damping materials, either in tension and compression or in shear, provides the mechanism for energy dissipation. Intimate contact between the material and the structure to be damped is imperative.

Generally, the materials must be applied in a layer approximately equal in thickness or weight to that of the panel being damped if appreciable damping is required. Usually the optimum location for applied damping is at areas of maximum movement.

Most efficient use of the damping material occurs if it is used as a "constrained layer"; that is, the material is sandwiched between the panel to be damped and a relatively rigid layer, such as thin sheet metal, above. This forces the damping material into shear as the panel vibrates, dissipating substantially more energy than when the material acts simply in extension and compression.

Formulation of damping materials is a relatively complex process. Good and poor materials may be very similar in composition.

Often the plastic or elastomer is loaded with heavy, inert, granular materials such as lead powder or galena to increase weight, or with mica to introduce Coulomb damping by means of friction between the particles suspended in the binder. However, simply adding such materials randomly may prove very disappointing. Usually it is best that the fillers be of such particle size and type that they actually rub together within the binder matrix.

A fair amount of damping can be accomplished by filling the space between surfaces or faces of a wall with a heavy, granular material, such as sand. When poured into the voids of a masonry block wall, for example, sand provides appreciable damping with a noticeable increase in Sound Transmission Loss. However, sand has limited utility in most damping applications, and it is often a nuisance to use.

Carpets and carpet pads, cushions, and various "lossy" blanket materials are also effective shock absorbers and panel dampers. Some acoustical absorbent materials, if intimately adhered to panels, are reasonably good, too.

There are many means of measuring and expressing damping effectiveness. Some of the more common criteria follow.

Decay Rate in dB/sec. This criterion is perhaps the most useful in evaluating subjective improvement. The "ring" of panels is eliminated, radiation from vibrating surfaces is reduced, and other noticeable effects are mitigated by damping. The effect is equivalent to reducing reverberation in a room.

The "Geiger Thick-Plate Test" procedure is a simple means of measuring decay rate. The method lends itself to quick comparisons between materials and is well-suited for research and development work and quality control.

Per Cent of Critical Damping. "Critical" damping is the damping necessary to just prevent oscillation. The damping ratio,

$$\frac{C}{C_c}$$

where:

C_c = the critical damping coefficient;
C = the damping coefficient of the material under consideration;

is one means of expressing this criterion.

The values themselves are not very informative unless compared with or calibrated against some other criteria. However, they are very useful in comparing the performance of products or in predicting performance of damped structures.

Loss Factor η. Damping is also often expressed as the loss factor, η, where:

$$\eta = \frac{1}{2\pi} \frac{D_o}{W_o}$$

where:

D_o = total energy dissipated in the system as the result of damping during one cycle;
W_o = total energy of vibration of the system.

η is related to the damping ratio (see above) as follows:

$$\frac{C}{C_c} = \tfrac{1}{2}\eta$$

η is also related to the resonance sharpness, Q, of the system as follows:

$$Q = \frac{1}{\eta}$$

where:

$$Q = \frac{f_o}{\Delta f_o}$$

and

f_o = frequency of resonance;
Δf_o = number of cycles between the two frequencies at either side of resonance where the Intensity Level is 3 dB down.

η is related to the decay rate of a free vibration as follows:

$$\Delta t = 27.3\eta f$$

where:

$$\Delta t = \text{decay rate in dB/sec};$$
$$f = \text{frequency in cps (Hz)}.$$

There are many means of measuring these various parameters. It is important only that the equipment used for dynamic measurements be sensitive enough to measure accurately without affecting appreciably the motion of the system. Various test arrangements are described in the References and Bibliography.

It is very important that tests be conducted on the materials to determine their damping characteristics. Often, apparently very similar materials vary widely in performance, particularly with changes in temperature and frequency.

Some materials depend upon combinations of viscous, Coulomb, and hysteresis damping. For most applications this is desirable, but for some applications possibly only one of the damping mechanisms is desirable.

Useful damping materials exhibit decay rates from as low as 5 dB/sec to over 80 dB/sec, and from $\frac{1}{2}\%$ to 20% of critical damping.

Damping ratios range from a fraction of 1% to over 20% of critical damping.

Sound Reinforcement

Sound control does not always mean eliminating or reducing sound. An important part of acoustical design involves the amplification or reinforcement of sound as well.

Wanted sound—the speech, music, and other signals which are desirable for communication, entertainment, and the like—must be presented at a level high enough to be heard clearly above the ever-present background noise, whatever its level may be.

The air within a space is usually the transmission path between the source and the listener. Because of its low impedance, air cannot accept acoustic energy at a high rate—it is an inefficient coupling medium between source and receiver. Sources small in area cannot "drive" much energy into the air, even though their energy input may be large.

As discussed earlier ("Shape and Configuration of Spaces," Section II, page 27) some spaces are so large that an unassisted source cannot develop enough acoustic energy to provide adequate sound level throughout the spaces. Fortunately, there are techniques for amplifying or reinforcing the source appreciably.

The Mechanism of Sound Reinforcement   There are two forms of *"natural" amplification*, both so old and commonplace that we often ignore them:

1. Using reflective surfaces to focus and direct as much as possible of the source acoustic energy output to the receiver.
2. Using a "horn" to improve the impedance match between the source and

the air, concentrating the energy as much as possible and directing it toward the receiver.

The first method is seen in the band shell, orchestra enclosure, and the pulpit canopy. They conserve and redirect to the listener energy which would otherwise be lost to the fly gallery or other areas.

The second method is seen in the megaphone (some readers may even remember the old trumpet horn used on early radio speakers and the primitive phonograph). A variation of this method often occurs inadvertently. If a small vibrating source (such as an electric shaver) is laid on a large, flat surface, such as a counter top, the sound of the source may be "amplified" considerably. It is important to remember that there is nothing mysterious or "magic" about the process; no laws of physics have been repealed, and no energy has been "created." All that happens is that more of the vibratory energy of the source is driven into the air, and more acoustic energy is transmitted via the air to the listener. The mechanical impedance "match" between the shaver and the counter top is good, and the large counter top is in contact with much more air than the small source, so it can "drive" more energy into the air.

Today, however, deliberate *"electrical" amplification* is the more significant means of sound reinforcement. This process involves a multiple energy conversion procedure:

1. First, the sound energy of the source drives the moving element of a microphone; the moving element modulates an electrical or magnetic field which modulates the electrical current which the amplifier adds to the circuit.

2. Then the modulated, amplified current drives a moving coil which is attached to a large diaphragm (the speaker cone) or to a small diaphragm at one end of an exponential horn (a sophisticated megaphone). The moving diaphragm moves a large volume of air at a greater amplitude than the original source moves it. This latter step converts the electrical energy back to the mechanical energy which we hear as sound.

Again, no energy has been "created." Some electrical energy has been added to the circuit and converted to mechanical (acoustic) energy.

Performance of Sound Reinforcement Systems   Reflectors and other "natural" reinforcement devices can rarely provide more than 3 to 6 dB "gain" over the unreinforced sound level. Yet this often means the difference between marginal and good listening conditions. When more gain is required, electrical amplification is necessary.

Understandably, most architects and many clients wince when the term "P.A. system" is mentioned. The miserable squawks and hollow booms coming from many systems are not conducive to delight or confidence when the need for sound reinforcement arises. A good sound reinforcement system is *not* a "P.A." system; the "voice of doom" sound, so often associated with warnings or unpleasant announcements at large gatherings, is an anachronism which does not belong in a pleasant acoustical environment.

The ultimate test of a good system is that listeners *are almost unaware* of its operation; they should be conscious only of adequate level, high intelligibility, and natural sound. In most cases, this is feasible with a well-designed system.

A sound reinforcement system is neither an incidental accessory nor a collection of hardware reluctantly added to a building because it can't be avoided. Whenever a system is required at all, it is an essential element in the acoustical design of the space in which it is used. It should be included in the design considerations from the beginning, and it should be designed as carefully as any part of the space. In fact, the sound system *is* a part of the space. Its operation is dependent upon and intimately connected with the form and function of the space of which it is a part.

A sound system will rarely overcome or correct serious architectural design deficiencies; rather, it may amplify and exaggerate the deficiencies. If there are echoes and excessive reverberation, the system will tend to reproduce and emphasize these shortcomings. If the room is too "dead" or "dry," only a very complicated and expensive installation can add any appreciable "liveness" or "warmth" to the space.

The system may be used for "full-range" reinforcement of all frequencies or merely to boost the level of "speech frequencies." It may only supplement natural sources or it may be the sole sound source, as for movies and some audio-visual presentations.

The main function of a reinforcement system is to increase "signal level" and to direct sound to listeners at a level high enough to be intelligible above the background. Yet there is a realistic limit to what can be accomplished in this way. It has been determined that the ideal amplification for most purposes is the 1/1 ratio—that is, the sound level at the listener's ear should be the same as the level about 36 in (1 m) away from the source (or at the normal distance from source to receiver). Thus, if the Sound Pressure Level measured at about 36 in (1 m) in front of a lecturer is about 65 to 70 dB, the level at the listener's position, anywhere in the hall, should be about the same or a trifle lower—but *not* higher. If this condition can be provided, the listener is often unaware of the fact that the sound is amplified at all (if other parameters, such as fidelity, direction, etc., are realistic and proper).

For adequate intelligibility, however, the signal should be at least 10 dB above the background noise. Therefore, if the background level is too high, it may be impossible to maintain the 1/1 ratio between the unamplified source level and the amplified level at the listener's position; it may be necessary to "outshout" the noise by raising the amplified level at the listener well above the original source level. Up to a point, this may make it possible to hear the signal, but it will be an obviously "amplified" signal, and listening to it may become tiring and even unpleasant. Certainly, if the amplified signal level at the listener's ear approaches 85 dB, it will soon become annoying and uncomfortable.

It has also been found that about 9 dB of amplification (increase in Sound Pressure Level over original source level, measured the same distance from the original source as from the sound system speaker) is all that can go "undetected"; more amplification invariably becomes obvious to the listener. In many noisy spaces, such as a sports arena, an airport waiting room, or similar areas, much more than 9 dB increase must be provided, so it is not possible to keep the listener unaware of the amplification; he will be quite conscious of it.

A top-quality sound system is capable of providing levels as high as 100 dB at a distance of 100 ft (30 m) from the speakers, when necessary. Further, it can be designed and adjusted to provide as much as 25 dB gain in a space before the obnoxious howling called *"feedback"* occurs. Feedback is perhaps one of the most undesirable by-products of electrical amplification. In some spaces, with poorly designed systems, it can occur even before measurable gain has been provided. However, it is not an inevitable concomitant of amplification; rather, it is evidence of an improperly designed system, probably operating in an acoustically substandard space.

The amplified sound must be a faithful reproduction of the original, or the "phony," distorted quality of the sound may become objectionable.

Directional realism must also be maintained. There are few things more disconcerting than to see the lecturer's lips move at the podium while the sound comes to the listener from behind, overhead, or from the sidewalls of the room. We are not as sensitive to vertical alignment of sound sources as to horizontal alignment; thus, the sound system speaker may be somewhat above the original source, but it cannot be far to either side without its becoming conspicuously "misplaced."

To reiterate, a "good" sound system, properly designed, properly installed, and properly adjusted, can be a most valuable and essential part of a good acoustical environment; a poor system can be a disaster.

Sound Reinforcement Equipment   The design of reflectors has been discussed elsewhere (Section II, page 29, see also Section III, page 138). A discussion of electrical sound reinforcement equipment and systems follows.

Each element of a system must be of the best quality that the client can afford, and it must be chosen correctly and used properly. Sound systems are "cheap" or good; there are really no acceptable compromises. Usually the best possible system is minimal; compromises with quality are invariably a mistake.

A good first rule in system design is to keep the system as simple as possible, using a few components and as little equipment as feasible. Every component is potentially a service and maintenance problem; the more components, the more likelihood of breakdown at a critical time.

The three principal elements of a system are:

1. The input (microphone, record player, tape deck, radio tuner, projector sound track, etc.)
2. The amplifier, preamplifiers, line amplifiers, controls, etc.
3. The output—usually speakers (although recorders and radio transmission lines are often the output end of some systems).

Systems are often identified as "low-level" (actually, "distributed speaker" system would be more accurate), with a number of speakers located near the listeners; and "high-level" (actually, "central speaker" system would be better), with a single speaker cluster located near the actual sound source. As the identifying terms imply, usually the distributed speaker system operates each speaker at a lower Sound Pressure Level than the somewhat higher level at which a central speaker system usually operates.

In addition to these two basic types of systems, various "hybrids" are occasionally found, including combinations of distributed and central speaker systems, "split" systems with one portion of the seating area covered from one speaker location and other portions covered from different speaker locations, etc. Occasionally, such systems are justified, but their use implies either an extraordinary space or a poorly designed space. Only a very skilled and competent practitioner should undertake the design of such systems, since there are many pitfalls in their design and use.

Ideally, a sound system should be designed from the speakers back to the inputs. Since the speaker (or speakers) are usually intended to supplement the sound source, *they should be located as near as possible to that source*. An exception to this rule is the case of audio-visual presentations, where the sound should usually appear to be coming from the visual image.

The speakers should direct their sound to the audience, not against reflective room surfaces and not up into unoccupied volumes of the space; and the speakers should "look into" absorptive areas. This usually means that rear walls and other room surfaces facing the speakers must be made absorptive, or serious echoes may result.

Accurate directionality and control of the speaker dispersion pattern cannot be accomplished for frequencies under about 250 cps (Hz); these long wavelengths diffract readily around even large objects; and the "horns" in which low-frequency cone-type speakers are located are often enormous boxes enclosing 8 ft^3 (0.25 m^3) or more volume. Their dispersion pattern is usually considered as about a 90-degree cone.

Remarkably good control of frequencies above 500 cps (Hz) is possible with properly designed horns. High-frequency "drivers" are connected to variously

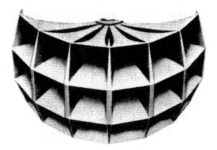

Figure 30 Large multicell horn. (Courtesy Altec Lansing, a Division of LTV Ling Altec, Inc.)

shaped horns, ranging from true exponential "trumpets" to radial expansion and reentrant horns of various types. The best and most accurate horn remains the "multicell" type (see Figure 30), consisting of two to eighteen exponential trumpets joined at the small end to an accurately shaped throat, attached to the driver (voice coil and diaphragm).

These units are capable of extraordinarily accurate control of the coverage pattern, even out to more than 100 ft (30 m) from the horn, for all frequencies above about 400 cps (Hz), and for levels up to more than 100 dB.

The speaker array for good, high-level *central systems* is obviously large and conspicuous (a multicell horn, alone, may be 42 in [1.1 m] deep, 26 in [0.67 m] high, and 38 in [1 m] wide) and not easily "tucked away" in some cornice or hidden space in a room. Further, it is not possible to "bury" the speakers behind some surface and let their sound "escape" through a small opening. For proper "coverage" of the audience, it is imperative that every listener in the space be able to see directly into the speaker (or that he would be able to do so if any decorative grille or covering were removed). Likewise the speaker must have an unobstructed line to each seat which it serves. Nothing must be inserted between speaker and listener except very porous "grille cloth" or very open grillework (about 70% open area is desirable) consisting of members not more than $\frac{5}{8}$ in. (1.5 cm) in width (the dimension of the member at right angles to the line from speaker to listener). Low-frequency sound will diffract around fairly large obstructions, but high-frequency sound is "cut off" by even small strips, slats, or framing members.

Distributed speaker systems usually employ many cone-type speakers, located near the listeners; hence they may be smaller, and small grilles may be used over each of them. Nevertheless, for good, full-range, high-fidelity reproduction of sound, the speakers must be 8 in (20 cm) or more in diameter (for speech reproduction only they may be as small as 4 in [10 cm] in diameter). No attempt is made to preserve the illusion of directional realism. The coverage pattern for each speaker can be considered a 60-degree to 90-degree cone; hence, many speakers are required for complete, uniform coverage in low-ceilinged spaces or where listeners are near the speakers. The most serious and frequent mistake in the design of distributed speaker systems is the use of too few speakers. When this is done, the level varies widely from location to location, and the resultant effect is one of "spotty" coverage. A good rule of thumb for speaker spacing in such systems is:

$$D = 2 (H - 4)$$

where:

> D = distance in feet between centers of speakers;
> H = ceiling height in feet.

or:

$$D = 2 (H - 1)$$

where:

> D = distance in meters between centers of speakers;
> H = ceiling height in meters.

Thus, in a 12-ft (3.5-m) ceiling, distributed speakers must be located not more than 16 ft (5 m) on centers. Obviously, a large, low-ceilinged room will require many speakers for adequate coverage. Equally obvious is the fact that a *good* distributed speaker system will often cost far more than a good central system (contrary to widely held opinion).

Reinforced sound should arrive at the listener's ear slightly later than the direct (natural) sound, but *never* more than 45 msec later or it will sound like an echo. This means that path difference (speaker-to-listener compared with natural-source-to-listener) must be less than 50 ft (16 m), and preferably well under 40 ft (12 m). When it is impossible to locate speakers to accompish this, time delays must be introduced into the system to provide the necessary delay in the reinforced sound. Usually this involves a digital delay mechanism, with the time delay accomplished by sophisticated electronic means. Such work should be handled only by an experienced and competent professional.

Obviously, the distributed speaker system can rarely maintain the ideal time relationship discussed above. Usually, most of the speakers will be nearer the listeners than the natural source will be. Either this must be accepted, or a delay mechanism must be used to correct the timing disparity.

A common (and usually poor) arrangement is to use two widely spaced speaker arrays at the front of a room, with their dispersion patterns overlapping in various audience areas. Another equally serious mistake is a distributed speaker system with wall-mounted speakers facing one another from opposite sidewalls of the room. Both arrangements tend to produce unpleasant audible effects; and they are almost certain to create areas of destructive interference where their coverage patterns overlap. In fact, any arrangement in which the speaker coverage patterns cross one another or overlap appreciably is likely to be unsatisfactory.

The *electronic amplification components* of a system must be as distortion-free as possible, and with sufficient power to bring the signal up to satisfactory levels while working well within their rated capacities. Usually, preamplification stages are required to boost the input signal to a level where the power amplifier can increase it to adequate levels for the speaker voice coils. Occasionally, line amplifiers are also used to increase the input signal level even further before it reaches the power amplifier. The type of input, length of microphone lines, and other specific job requirements determine how much and how many stages of amplification are necessary. It is important that adequate amplification is available without overworking the equipment. It is imperative also to know what the power rating of amplifiers actually means; some are rated at "...x-watts *peak* power...," which is a means of seriously overrating the equipment. It means that for brief pulses, such as occasional musical notes or shouts, the amplifier can handle the load, but for continuous loads, the equipment can handle about half the rated "peak" power.

Today, most equipment is "solid state." Some of the better, newer solid-state power amplifiers have built-in overload protection; they appear to function satisfactorily even under heavy loads.

Transformers and impedance-matching devices, while small and apparently insignificant, can make the difference between a good and a mediocre or poor installation. It is folly to use top-quality electronic components and cripple their performance with cheap, high-loss transformers or other questionable "economies."

Microphones of many types and performance characteristics are available today, ranging from the superb wide-range condenser types to cheap, distortion-producing units which have no place in a good sound-reinforcement system. Their distinguishing characteristics include their pattern (cardioid, omnidirectional, etc.), frequency range, smoothness of response, sensitivity, and stability. Again, to compromise quality for a trivial cost saving is to risk compromising the quality of the entire system.

Ideally, the natural source should be as close to the microphone as practical, and a constant distance from the microphone, so that the signal-to-noise ratio is high and uniform. Few things are more disturbing than to have a sound system "blast" when the source comes too close to it, and "fade" as the source moves too far away. Therefore, placement of microphones and microphone outlets is an important part of good acoustical design in a space. In special areas, many microphones are located in an array which keeps the source within the field of a microphone and about the same distance from a microphone at all times. Conduit is cheap, so it is always wise to provide enough microphone outlets to handle at least the predictable situations; a few spare outlets are good, low-cost insurance against the unpredictable.

Microphone cable should always be top-quality, shielded cable, specifically designed for this use. *No splices* should ever be permitted between microphone and amplifier; if connections are required, they should be made at a terminal strip or in a junction box with a terminal strip. Speaker wire should be twisted pairs capable of handling their load for the distances of run with trivial loss; they should *never* be run in the same conduit with microphone lines, nor even very close to them. Again—absolutely no splices should be permitted. It is extremely difficult to run down line breaks at best, and splices are so frequently the cause of problems and breakdowns that they simply should not be permitted.

The control panel for *any* sound system should be in a convenient location where an operator can adjust it while looking at and listening directly to the speakers. Only thus can the operator know what the listening audience hears. Only in unusual situations should the system require the operator to preset levels and adjust them from a remote monitoring station, since he can never be sure that his monitoring indications truly reflect what the audience hears. Obviously, then, the control panel should be either at the rear of the room or in a position as remote from the speakers as are the more distant seats. If an enclosed booth is provided, a large operable window should be located in the *front* of the booth so that the operator can open it to permit listening directly to the speakers while operating the controls.

The essential controls should always include (among other things) level control for *each* input (microphone, tuner, tape deck, etc.) and *each* output. It is often necessary to adjust levels of various speakers to correct for varying room occupancies and varying or unpredictable background noise conditions. If this flexibility is missing from the system, its usefulness is seriously compromised.

As mentioned earlier, any room is a "musical instrument" which "tunes" to certain frequencies more strongly than to others. In almost all spaces, some of the room response modes are much stronger than the average response level of the room. When these modes are amplified, they may set the sound system into "feedback" (the howling, screeching sound which builds up until the system is shut off or turned down) long before adequate system gain has been achieved in the space. In fact, in highly reverberant spaces, some systems cannot provide any useful gain before feedback occurs. Even in well-designed spaces with moderate reverberation times, the gain before feedback may be too low to permit the sound system to provide adequate level in remote seats or in noisier areas within the space.

Fortunately, it is possible to adjust and modify *good* equipment to correct this problem. Simply adjusting the bass and treble tone controls may help considerably (particularly attenuating the bass portion of the spectrum); but for significant improvement, and for high gain in highly reverberant spaces, special *"equalization"* procedures are required. These techniques, in the hands of skilled technicians, can provide dramatic improvement in the performance of a system, even within "impossible" spaces. The techniques consist, essentially, of inserting carefully designed, narrow-band, adjustable filters into the circuitry of the system, so that room resonances are simply not amplified. Thus, the frequencies which excite the offending modes are not amplified (or are amplified so slightly that feedback does not occur). The entire procedure obviously requires the skill and training of experts but it is being used increasingly in the better large systems in critical installations where top-quality sound and high gain are imperative.

Obviously this brief discussion of sound reinforcement merely outlines the fundamental requirements for systems, so that the architect or engineer can plan intelligently for the use of such equipment in spaces which he designs. As in all of his work, the professional is dependent upon the competence of the designers, manufacturers, and installers of the equipment and materials used. However, rarely is the ultimate performance of equipment so obvious to everyone. Therefore, compromises in design or quality of sound systems will quickly become evident, even to the point of degrading an otherwise good space. (For detailed discussions of sound system equipment and design, refer to References and Bibliography, Section III, page 234.)

Mechanical Equipment Noise and Vibration Control

The mechanical and electrical equipment which is such an integral part of modern living is also the source of many acoustical problems. Intentional or otherwise, the choice of a heating, ventilating, or air-conditioning system, for example, is an acoustical decision. Rarely is the "acoustical equation"* so complete as in such systems. Nearly every piece of mechanical equipment in a building is a source of sound or vibration; the piping and duct systems are superb paths for both airborne and structure-borne sound; and the systems inevitably lead to some occupied space where people (the receivers) are usually present. Even within occupied spaces, an increasing amount of mechanical equipment is used for the normal activities conducted there.

Noise Sources   Nearly all equipment containing moving parts—whether rotating, reciprocating, vibrating, or impact-type—produces sound and vibration. Even well-balanced moving parts may, by their nature, produce sound. A fan, for example, even if perfectly balanced, will produce sound which is some harmonic of the number of its blades multiplied by the revolutions per second of the fan wheel. As each blade passes the "cutoff" of the scroll, a pressure pulse is

*Source–Path–Receiver.

created; the multiple pulses provide the alternating pressure in the air stream which, by definition, is sound. The same thing is true of centrifugal pumps, blowers, turbines, and similar equipment.

Reciprocating and internal combustion equipment produces sound which is usually a multiple of the piston movement.

The magnetostrictive effect in the laminated cores of transformers, ballasts, and similar electrical equipment produces sound which is usually some harmonic of 120 cps (Hz) (in the United States, where 60-cycle current is most common).

Tapping, hammering, punching, and similar actions produce impact sound and vibration which are normally "broadband" in frequency. Sliding, scraping, etc. (fundamentally high-speed "impact" in character) produce characteristic sounds, usually of distinctive frequency distribution.

The flow of fluids and gases in pipes and ducts is also a source of sound. If the flow were completely laminar and viscous, little sound would be created; however, at Reynolds numbers somewhere between 1300 and 2200, the flow changes from laminar to turbulent. The multiple vortices produced as the flow becomes turbulent are a source of "flow noise."

As flow velocity increases, the amount of acoustic energy produced increases dramatically. For example:

1. As gas velocity increases and approaches jet velocities, the sound power increases as the *eighth* power of the velocity.
2. Sound power output at a typical air-conditioning diffuser tends to increase as the *sixth* power of the cfm flow through the diffuser and as the *cube* of the pressure drop.
3. Sound power output at a restriction, such as dampers or throttling devices in the air stream, tends to increase as the *sixth* power of the amount of throttling and the *cube* of the increased resistance through the restriction.

Valves, orifices, and similar restrictions in fluid and gas lines are always sources of some sound; and, in some instances (steam-pressure-reducing valves, for example), they may produce very high sound levels.

Air and steam discharge through orifices always produces sound, often with considerable high-frequency energy content.

(See Section III, page 94, for a list of typical noise sources, and page 158, for the effect of operating parameters on equipment noise output.)

Transmission Paths   The sound may be airborne or structure-borne or both, in most installations. It may travel through the air in the building, through the building structure, and down the walls of pipes and ducts and similar paths. In addition, the acoustic energy may travel via the fluids and gases within the pipes and ducts. This is particularly true of the steam, water, and air in the typical plumbing, heating, and ventilating systems in modern buildings.

Noise and Vibration Control Measures   Noise and vibration control for mechanical equipment and systems involves the same techniques and principles discussed previously. Either the sound must be controlled at the source, the path must be interrupted, or the receiver must be protected. The acoustic energy may be absorbed or contained, the paths may be cut or modified by means of resilient connections, or the receivers may be isolated or protected by any of several methods.

Noise control at the source is ordinarily the most economical and efficient approach. For example, well-designed fans may be more than 10 dB quieter than other types of the same capacity. Careful design will eliminate a lot of noise in gears and bearings. As in all phases of engineering, there is no substitute for good design.

Properly balanced moving parts, smooth, well-lubricated bearings, and well-maintained machinery produce much less noise and vibration, so there is less unwanted energy to deal with.

Low flow velocities in ducts and pipes, a minimum of throttling devices or abrupt changes in cross-section, and good aerodynamic design of all elements in a system tend to minimize acoustical problems.

Any change in a process or procedure which eliminates noise is invariably more effective and economical than noise control techniques to get rid of acoustic energy already produced.

While the actual acoustic energy associated with noise and vibration problems is normally quite small, noise is frequently an indicator of poor mechanical condition of the equipment and of unsatisfactory operation.

Isolating the entire vibrating assembly from its supports will reduce energy transmission to the supporting structure. This will minimize the dynamic effect of the vibration on the supporting structure, reduce the vibratory energy which might otherwise be felt by the occupants of the surrounding space, and minimize radiation of sound from large surfaces attached to the structure. Radiation from attached surfaces can be further reduced by applying damping materials to them. (A discussion of resilient mounting techniques and damping materials may be found in preceding portions of Section II, and in Section III.)

Enclosing the offending equipment within partial or complete barriers is often successful in reducing noise (*if* the equipment is not rigidly attached to the floor or structure or otherwise "short-circuiting" the enclosure).

Occasionally the enclosure is set into vibration, and it, too, must be mounted resiliently to isolate its vibration from the floor or building structure.

Connecting pipes, ducts, and conduits may act as unwanted transmission paths unless they are broken with flexible or resilient connections near the machine. It is often necessary to suspend such connecting lines with resilient supporting devices for some distance from the machine to prevent excessive transmission into the supporting structure.

Absorption alone provides little help in machine quieting. Absorption within an enclosure may be reasonably helpful, but small amounts of absorbents merely located near the offending machine are almost useless. Absorbents located on reflecting surfaces (of partial barriers, the ceiling or walls of the building, etc.) will minimize reflections and will reduce the reverberation within the space; however, in practice only a few dB reduction can be expected from this approach. The room may become more comfortable, and the machine operator may be able to localize and identify better sounds, particularly from his own machine; but noise levels usually will be reduced very little. Still, in designing a new building to house noisy machines, it is usually advisable to use absorptive roof decks (or ceilings), wall materials, etc., since the improved acoustical conditions usually justify the small added cost of absorption designed into the structure.

Absorbent linings within air ducts and special absorbent "mufflers" can be very effective in minimizing noise transmission down the length of the duct. (See Section III, page 157 for a discussion of various noise and vibration control techniques for typical installations.)

Shock Isolation "Shock" is an inexact term, usually used to describe rapidly applied impulsive force or energy, such as hammer blows or the impact of a dropped object. The energy is applied quickly, and the motion is stopped quickly.

(**Note:** The term "jerk," meaning the rate of change of acceleration, is occasionally used in discussions of shock and shock isolation.)

Vibration isolation devices usually are lightly damped; they store most of the energy which they accept from the driving source, and they return it to the source. Shock isolation usually involves means of slowing the rate of application of the impulsive energy to the base, high damping, and dissipation of much of the energy as heat within the shock mounts; little energy (if any) is returned to the source. Shock isolation is a complex procedure, and usually should be undertaken only by experts.

Sound Fields and Sound in Enclosures

Increasingly, the designer must be able to predict the acoustical effect of equipment operation, both within the room in which it is located and the out-of-doors to which the sound is transmitted. For example, the effect of locating an air-cooled condenser on the roof, adjacent to a neighboring apartment; or the effect on workers within a plant of installing a large exhaust fan, a diesel engine-generator set, or a large new milling machine must be forecast with reasonable accuracy if the designer is not to run afoul of neighborhood zoning ordinances or OSHA requirements. Some simple mathematics for calculating the source–path–receiver equation follows.

Point Source If we could conceive of all the acoustic energy of a source originating at a point and radiating spherically, we could, with simple mathematics, define the sound field around that point.

For example, we would immediately see that the radiated energy at any distance from that point would vary according to the inverse square law. Thus, the intensity level would drop by 6 dB as the distance from the source doubled, since the same amount of energy is distributed across four times the area. (Surface area of a sphere varies as the square of the radius.)

Sound Power Therefore, if we measured the sound pressure level anywhere on the surface of a sphere of 1 m^2 surface area around a point source, we would have the Sound Power Level of the source (by definition).

Line Source If we had a line of closely spaced point sources or a line of steadily moving sources (such as steadily moving traffic), radiation from that line would tend to be cylindrical. The intensity level would vary with the radius, and the level would drop by 3 dB as the distance from the line source doubled. (Surface area of a cylinder varies directly with the radius.)

Random Sources Most noise sources tend to be somewhat random, rather than point or line sources. In fact, few sources can be considered point sources unless they are very small and observations are made at a considerable distance from them. Large machines, for example, are often a combination of point sources, large vibrating panels, and possible line sources. Therefore, calculation of the sound field radiated from them is rarely a simple, straightforward matter.

Directivity Few sound sources are located in free space, radiating uniformly in all directions. A point source, on a hard, reflective surface, for example, radiates most of its energy through the hemisphere above the hard surface. Thus, its energy would radiate through half of a sphere, and the intensity anywhere on the surface of that hemisphere would be twice what it would have been had it been radiating spherically. The source would be said to have a *directivity* of 2.

A source located at the intersection of two planes at right angles to each other would have a directivity of 4 (all the energy radiates through one-fourth of a sphere); at the intersection of three planes at right angles to each other the point source would have a directivity of 8 (all the energy radiates through one-eighth of a sphere); etc. Directivity is usually noted as Q in acoustical equations.

Free Fields If the energy from a source radiated into a "free" (totally absorbent and nonreflective) field, there would be no energy "build-up." The sound energy density and pressure level would be dependent solely upon the source level and the distance from the source.

Thus, if we knew the Sound Pressure Level at any given distance from a point source, we could determine the *Sound Power Level* of that source. Conversely, knowing the Sound Power level, we could calculate the Sound Pressure Level.

$$\text{Sound Power Level} = SPL + 20 \log_{10} r + 0.7 \text{ dB}$$

where

> SPL = the Sound Pressure Level in dB;
> r = distance from point source to measurement position in feet;

or:

$$\text{Sound Power Level} = SPL + 20 \log_{10} r + 11 \text{ dB}$$

where

> SPL = the Sound Pressure Level in dB;
> r = distance from point source to measurement position in meters.

When we know the directivity of the source and its Sound Power Level, we can calculate the Sound Pressure Level anywhere within the field by the following equation:

$$SPL = PWL + 10 \log_{10} Q - 20 \log_{10} r - 0.7 \text{ dB}$$

where

> SPL = the Sound Pressure Level in dB;
> PWL = the Sound Power Level in dB;
> r = distance from point source to measurement position in feet;
> Q = directivity;

or:

$$SPL = PWL + 10 \log_{10} Q - 20 \log_{10} r - 11 \text{ dB}$$

where

> SPL = the Sound Pressure Level in dB;
> PWL = the Sound Power Level in dB;
> r = distance from point source to measurement position in meters;
> Q = directivity.

(Note: SPL is often written as L_P; PWL is often written as L_W.)

Reverberant Fields If the sound field around a source is enclosed or contained within surfaces, the energy will not dissipate freely, but some of it will be reflected and contained within the sound field. As a result, there will be a "build-up" of energy within the field. Furthermore, rarely will the field be uniform and "diffuse." Rather, there will likely be "room modes" and "standing waves" determined by the shape and dimensions of the room or enclosed space and the absorptivity of the surfaces.

Knowing the characteristics of the enclosure and the sound source, we can calculate the Sound Pressure Level anywhere within the enclosure:

$$SPL = PWL + 10 \log_{10} \left[\frac{Q}{4\pi r^2} + \frac{4}{R} \right] + 10.5 \text{ dB}$$

where

r = dimensions in feet;

$R = \dfrac{\alpha S}{1 - \alpha}$;

S = total area of the room surfaces;

α = average absorption coefficient of the surfaces at a given frequency;

or:

$$SPL = PWL + 10 \log_{10} \left[\frac{Q}{4\pi r^2} + \frac{4}{R} \right] + 0.2 \text{ dB}$$

where

r = dimensions in meters;

$R = \dfrac{\alpha S}{1 - \alpha}$;

S = total area of the room surfaces;

α = average absorption coefficient of the surfaces at a given frequency.

(See Section III, page 184, for further applications of these calculation procedures for rooms and enclosures.)

Section III

DATA

CRITERIA

DESIGN PROCEDURES

TROUBLE-SHOOTING

GLOSSARY

BIBLIOGRAPHY AND REFERENCES

Section III is intended to be a quick-reference guide and a practical handbook for the busy professional. The information included is sufficient to cover most of the more common, everyday problems encountered on the usual, typical projects. However, like any handbook or reference guide, its usefulness is dependent to a large degree on the reader's understanding of the principles which lie behind the procedures and practices. Therefore, we strongly urge that the reader at least read through (and preferably become familiar with) the preceding sections. As discussed in the Preface, this book was written with a particular purpose; an understanding of what is included in Sections I and II is essential to that purpose.

Several formulas, figures, and tables from Sections I and II have been repeated in Section III to avoid unnecessary cross-references.

Design Procedure

1. Determine sound and noise levels. 85
2. Establish sound and noise criteria 88
3. Site planning 91
4. Layout and orientation of rooms. 93
5. Choose exterior construction 95
6. Choose interior construction. 96

   a. Walls, floors, partitions 98
   b. Floor, wall, and ceiling finishes 133
   c. Shape and configuration 137

7. Sound reinforcement. 148
8. Mechanical, electrical, plumbing noise 153
9. Sound fields and sound in enclosures 184

An acoustical analysis consists of a series of logical steps:

1. Determining the use of the structure—the subjective needs.
2. Establishing the desirable acoustical environment in each usable area.
3. Determining noise sources inside and outside the structure.
4. Studying the location and orientation of the structure and its interior spaces with regard to noise and noise sources.

Acoustical design consists of two steps:

1. Designing shapes, areas, volumes, and surfaces to accomplish what the analysis indicates.
2. Choosing materials, systems, and constructions to achieve the desired result.

TABLE 1   Velocity of Sound in Various Media

Material	Approximate Sound Velocity (ft/sec)	(m/sec)
Air	1,100	335
Wood	11,000	3,350
Water	4,500	1,370
Aluminum	16,000	4,880
Steel	16,000	4,880
Lead	4,000	1,220

It is usually pointless to dispute changes or differences of less than the amount necessary to make a "just perceptible" change. A tolerance of $\pm 2\frac{1}{2}$ dB is usually acceptable in identifying levels or in meeting specifications. In general, unless a level or criterion by definition must not be *exceeded*, any number or level or class implies a tolerance of not more than $\pm 2\frac{1}{2}$ dB.

Historically, in practice, acoustical ratings and groupings or "classes" of materials have tended to fall into 5-point ranges, such as STC-37, STC-42, STC-47, etc.; Noise Criteria are often given as NC-30 to NC-35, etc.; and Sound

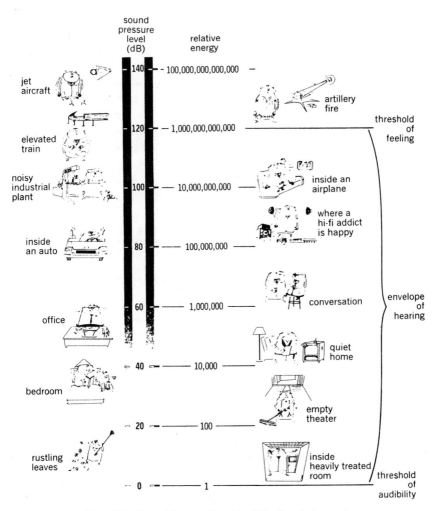

Figure 3-1 Sound Pressure Level in dBA of typical sounds.

TABLE 2 Subjective Effect of Changes in Sound Characteristics

Change in Energy Level	Change in Sound Level	Change in Apparent Loudness
26%	1 dB	Insignificant
Doubling	3 dB	Just perceptible
Tripling	5 dB	Clearly noticeable
Ten Times	10 dB	Twice as loud (or $\frac{1}{2}$)
100 Times	20 dB	Much louder (or quieter)

Absorption Coefficients are usually shown as .60, .65, etc. (although 10-point groupings are often used for these latter ratings).

From Table 2 it can be seen that a 5-dB difference is clearly noticeable, but $2\frac{1}{2}$ dB is hardly perceptible.

Unless indicated otherwise, this book uses the A-weighting of the standard sound level meter to identify overall Sound Pressure Levels by a single number (as indicated in Figure 3-1). Where a spectrum analysis is indicated and values are given in octave bands (or fractional-octave bands), the Sound Pressure Level is always the actual unweighted value in the band (using the "flat" weighting net-

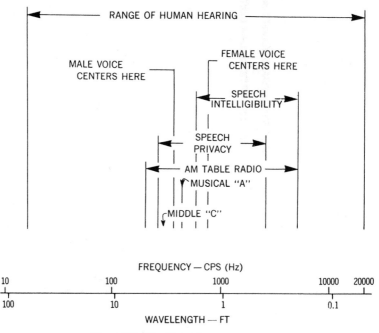

Figure 3-2 Significant frequency ranges.

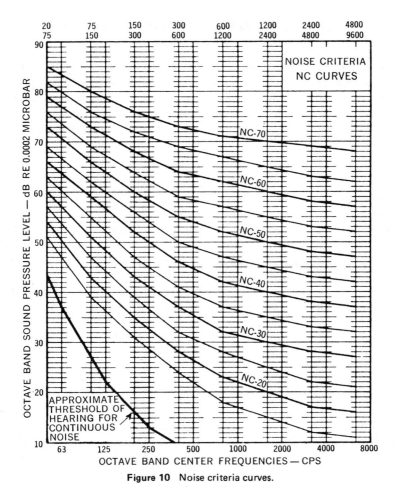

Figure 10 Noise criteria curves.

work of the meter). Section II, page 19, indicates the rationale and background for this procedure.

When an "NC-number" is used to specify an environment, it means that in no frequency band shall the Sound Pressure Level within the space (measured in the region of normal human activity within the space) exceed the specified NC-Curve (Figure 10) (indicated by NC-number).

In practice, a tolerance of $\pm 2\frac{1}{2}$ dB is implied when an NC-number is used to describe an environment. In specifications, an allowance of about 2 dB above the curve is usually accepted in any one octave band.

Background Noise Criteria

NC-levels usually refer to steady, continual *background* levels within a space or neighborhood, as opposed to specific noises or intermittent activities occurring there. The level of a performing orchestra or band, for example, normally is not expressed in this way, but in dBA or Sound Pressure Levels in various frequency bands. All numbers listed may vary as much as ±5 points in specific cases.

TABLE 3-1 Spaces

Subjective Classification	Function	Space	NC
	Testing, audiometry	Anechoic rooms	<15
	Recording	Voice studio	20
		Music studio	20–25
Quiet	Broadcasting	Radio or TV studio	20–30
	Sleeping	Bedrooms	30
		Hospital rooms	30
	Study	Library	30
	Drama	Legitimate theater	25
		School auditoriums	<30
		Motion pictures	30
Critical hearing and listening	Music	Concert and recital halls	25–30
	Speech	Lecture halls	<30
		Assembly halls	30
		Courtrooms	30
	Worship	Churches	30
	Discussion	Seminar rooms	30
		Classrooms	30
		Conference rooms	25–35
	Mental and creative tasks	Doctor's offices	30
		Executive offices	30
Normal		Normal private offices	35
		Study rooms	35
		Laboratories	35–40
	Public spaces	Corridors	40
		Lobbies	40
		Rest rooms	40

TABLE 3-1 (continued)

Subjective Classification	Function	Space	NC
Noisy	Dining	Restaurants	45
		Kitchens	55
	Clerical	Stenography and duplicating	50
		Banking floors	50
	Shopping	Stores, shops, supermarkets	40–50
Very noisy	Sports	Arenas	45–50
		Stadiums	55
	Transportation	Airports	55–65
		Railroad stations	55–65
		Garages and parking ramps	55–65
	Computing and calculating	Accounting rooms	65–70
		Computer rooms	70
	Machine rooms	Building mechanical equipment	70
	Production	Factories and shops	50–75

TABLE 3-2a Neighborhoods

Subjective Classification	Potential Exposure	Approx. NC-level
Quiet	Suburban and rural; residential; country or golf course areas	30–35
	Housing development	40
	Road—lightly traveled, low speed, no trucks	50
Moderately noisy	Shopping center, with parking lot	65
	Schools, with playgrounds; swimming pools	65
	Sports stadiums, arenas	70
Noisy	Industrial areas; power substations; pumping stations	70–80
	Highways and expressways; railroads and switchyards	75
	Airport aprons and runways	>75

Other Criteria

Other commonly used terms to describe criteria include:

TABLE 3-2b **Other Criteria**

NC or Noise Criterion	dBA minus 7 to 10 dB
PNdB or Perceived Noisiness, usually of aircraft sounds	dBA plus 13 to 15 dB
L_{eq}	Uniform dBA level which would contain the same total energy as a varying level
L_{dn}	L_{eq} with 10 dB additional weighting for nighttime hours
L_n	dBA level exceeded $n\%$ of time, such as L_{10}, L_{90}, etc.
L_{50}	dBA level exceeded 50% of time; approximately 3 dB less than L_{eq} for typical noises
NEF or Noise Exposure Forecast	Calculated value usually used to describe areas near airports
CNR or Composite Noise Rating	Calculated values, with special weighting for times, number of noise events, type of noise, etc.; usually in association with environmental noise

Note: Most acoustical criteria are constructs, designed to distill a number of variables and parameters down to a single value which may be used (with caution) to compare *similar* situations or phenomena. New criteria appear frequently, but dBA values, or actual SPL values at various frequencies, the character of the noise (steady, impulsive, repetitive, etc.), and the time of occurence are normally more informative.

Site Planning

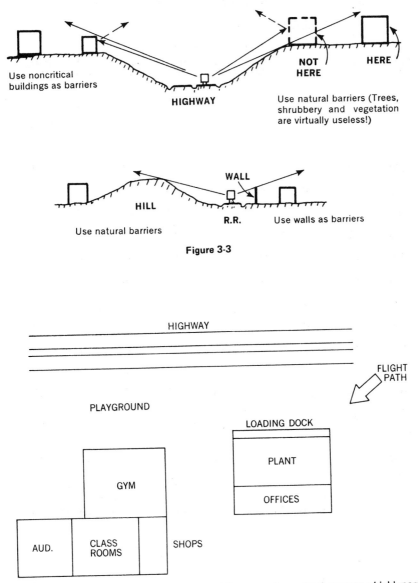

Use noncritical
buildings as barriers

HIGHWAY

NOT
HERE

HERE

Use natural barriers (Trees,
shrubbery and vegetation
are virtually useless!)

WALL

HILL

R.R.

Use walls as barriers

Use natural barriers

Figure 3-3

HIGHWAY

FLIGHT
PATH

PLAYGROUND

LOADING DOCK

PLANT

GYM

OFFICES

AUD.

CLASS
ROOMS

SHOPS

Figure 3-4 Orientation on the site. Orient building to use less critical spaces to shield more critical spaces. Put noisier activities on noisy exposures, quieter activities on quieter exposures. Higher-frequency noise (tire whine, etc.) is easily shielded against; lower frequencies (rumble and roar) tend to diffract around barriers.

TABLE 3-3　Effect of Distance

Reduction of Sound Pressure Level for Distances Beyond 50 ft.

Sound Pressure Level Reduction (dB)

Distance (ft)	Frequency Band (cps)							
	20 75	75 150	150 300	300 600	600 1200	1200 2400	2400 4800	4800 10000
50	0	0	0	0	0	0	0	0
63	2	2	2	2	2	2	2	2
80	4	4	4	4	4	4	4	4
100	6	6	6	6	6	6	6	6
125	8	8	8	8	8	8	8	8
160	10	10	10	10	10	10	10	10
200	12	12	12	12	12	12	12	12
250	14	14	14	14	14	14	14	14
320	16	16	16	16	16	16	16	16
400	18	18	18	18	18	18	18	18
500	20	20	20	20	20	20	20	21
630	22	22	22	22	22	22	22	24
800	24	24	24	24	24	25	26	29
1000	26	26	26	26	26	28	30	33
1250	28	28	28	28	28	31	33	38
1600	30	30	30	30	31	34	37	43
2000	32	32	32	33	35	38	42	49
2500	34	34	34	35	37	41	46	54

(See also Section II, page 80, and Section III, page 184, "Sound fields and sound in enclosures.")

Layout and Orientation of Spaces

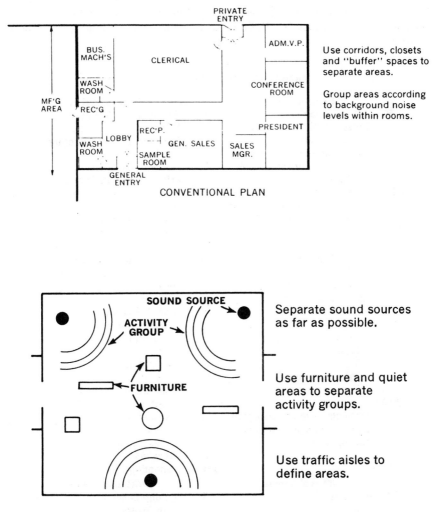

Use corridors, closets and "buffer" spaces to separate areas.

Group areas according to background noise levels within rooms.

CONVENTIONAL PLAN

Separate sound sources as far as possible.

Use furniture and quiet areas to separate activity groups.

Use traffic aisles to define areas.

OPEN PLAN

Figure 3-5 Space planning for acoustical privacy.

Typical Noise Sources

Typical noise sources, both exterior and interior, reach levels far in excess of the steady background noise levels listed in Tables 3-1 and 3-2. They may be sporadic and intermittent or frequent and continued. Some common sources and their approximate levels (in dBA) are listed below, ranked from least noisy to noisiest:

TABLE 3-4 Typical Noise Sources

Communication

60	Conversation
	Telephones
to	Intercom systems
	Paging and announcements
	Audio-visual presentations
85	Teletype
dBA	Bells, buzzers, horns*
	(*In very noisy spaces, their levels may exceed 100 dBA.)

Productive Activities

70	Typing and office machines
	Mail room
	Cleaning and service equipment
to	Vending machines
	Printing and duplicating
	Accounting and computer equipment
	Kitchen equipment
110	Construction and repair
dBA	Manufacturing equipment

Transportation

70	Walking
	Elevators
	Escalators
	Dollies and carts
to	Vehicles:†
	Autos
	Buses
	Trucks
	Railroad trains
120	Subway trains
dBA	Aircraft
	(†Usually exterior, except in garages, hangars, subway tubes, etc.)

Entertainment

70	Radio, television
	Voice
	Audiences
	Musical instruments
	Sound systems
	Circus and carnival
to	Sports:
	Weight lifting
	Boxing
	Swimming
	Football
	Basketball
	Hockey
	Roller skating
120	Bowling
dBA	Shooting

Miscellaneous

70	Animals
	Toys
to	Fireworks
120	Farm machinery
dBA	Military activities

HVAC Equipment

35	Lighting ballasts
	Air diffusers
	Window air conditioners
	Transformers
to	Plumbing
	Toilets
110	Pumps and fans
dBA	Chillers and compressors

Choosing the Exterior Construction

Whether you are trying to keep extraneous noise *out* or internal noise *in*, the exterior construction—*including the roof, windows, and doors*—must serve as a barrier to maintain the difference in Sound Pressure Levels between inside and outside.

If you can determine the outside levels, and you can establish acceptable inside levels, the exterior construction must provide enough isolation to maintain the difference in level between outside and inside (see Figure 11). (This is true, too, for partitions between rooms.)

Remember—ordinary glass is a rather poor sound barrier; louvers, holes, or openings of *any* kind are potentially serious sound "leaks." Lightweight curtain wall construction is subject to severe "flanking" transmission.

Don't forget noise such as rain on the roof, aircraft flyovers, roof-mounted mechanical equipment, walking on the roof, the big air intake opening for the supply fan located right under a first floor window, etc.

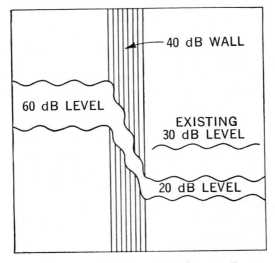

Figure 11   A wall as a "sound barrier."

Choosing the Interior Construction

The interior construction systems and materials—floors, walls, and ceilings—must serve as acoustical barriers, reflectors, diffusers, and absorbers, in addition to all of their other functions. None should be chosen until the acoustical requirements and criteria for each area have been determined.

If acoustical criteria can be established for each space, various materials and constructions may be considered according to their ability to fulfill the acoustical requirements. In the absence of actual measurements or specific criteria or requirements, the tables and charts in the following pages of this section give reasonably safe, general criteria.

(Refer to Section III, page 130, for a discussion of "Masking and Privacy.")

Performance characteristics of materials and constructions are listed in the following tables. For example, the STC-ratings for floors, walls, and partitions of almost any type used in normal construction in the United States (and many other countries) can be determined from the tables supplied; further, the designer may choose a basic type of construction (masonry, dry-wall, plaster, etc.) and, using the tables, decide how to bring its isolation performance up to almost any required ratings by various modifications.

Current absorption data for acoustical tiles and panels, ceiling attenuation factors (through-the-ceiling STC-ratings), absorption data for carpets and furnishings, etc., may be obtained from the current annual *Sweet's Catalogs* or directly from manufacturers or trade associations. Some general absorption data are also listed in this section and in Section II, page 52, of this book.

Performance of constructions as barriers to airborne sound can be described reasonably well with a one-number rating system. Actual tests are made at 16 frequencies, and the curve of the Transmission Loss is compared with a "standard contour" which reflects the characteristics of human hearing (see Figure 15). The Transmission Loss value where the contour intersects the 500 cps (Hz) ordinate is called the Sound Transmission Class (STC) of the construction. (Refer to Section II, page 38, for further discussion of STC.)

In the following tables, the STC-rating number indicates the approximate center of a range; for example:

> STC-37 means STC-35 to less than STC-40
> STC-42 means STC-40 to less than STC-45, etc.

All criteria are based upon normal background level in "Room Being Considered" and average construction cost range for the building.

For low-cost construction or higher than normal background levels, criteria may be reduced 3 to 5 points. For high-cost construction or lower than normal background levels, criteria may be increased 3 to 5 points.

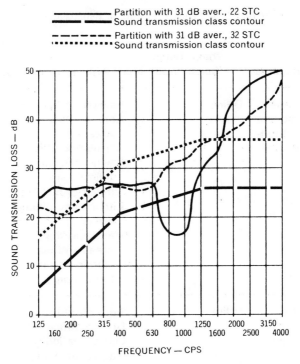

——————— Partition with 31 dB aver., 22 STC
━━━ ━━━ Sound transmission class contour
— — — — — Partition with 31 dB aver., 32 STC
▪▪▪▪▪▪▪▪▪▪▪ Sound transmission class contour

Figure 15   Determination of Sound Transmission Class.

All exterior or other areas which are the source of extraneous noise are assumed to have a normal or usual noise level and spectrum for such areas. If extraordinarily high noise levels exist (as in the vicinity of engine test stands, airports with heavy jet traffic, or heavily traveled highways with heavy truck traffic, etc.), increase all isolation requirements by up to 5 points, or have measurements made by a qualified acoustical consultant.

The tables provide a fairly reliable, conservative guide for choosing partition constructions. They assume a "normal" spectrum for the background sound in the room; no unusual or extreme extraneous noise conditions; and no "special" conditions or highly critical uses for the spaces.

For auditoriums, theaters, recording studios, music schools, and similar areas; for large hotels, apartments, dormitories, and similar buildings; and for any structure where the amount of wall construction is large enough to make the economic multiplier very large, professional advice is the only safe, economical course.

Floors (and roofs) must provide at least as much isolation (and preferably 3 to 5 points more) against *airborne* sound as walls or partitions surrounding the spaces.

TABLE 3-5a   Wall, Partition, or Panel Between

Type of Occupancy	Room Being Considered	and	Adjacent Area	Sound Isolation Requirement
Executive areas, doctors' suites, confidential privacy requirements	Office		Adjacent offices	STC-47
			General office areas	42
			Corridor or lobby	47
			Washrooms and toilet areas	47
			Exterior of building	42
			Kitchen and dining areas	47
			Manufacturing areas and mechanical equipment rooms	52
Normal office areas, normal privacy requirements	Office		Adjacent offices	STC-37
			General office areas	37
			Corridor or lobby	37
			Washrooms and toilet areas	42
			Exterior of building	37
			Kitchen and dining areas	42
			Manufacturing areas and mechanical equipment rooms	47
Any normal occupancy, using conference rooms for group meetings or discussions	Conference rooms		Other conference rooms	STC-42
			Adjacent offices	42
			General office areas	42
			Corridor or lobby	42
			Washrooms and toilet areas	47
			Exterior of building	37
			Kitchen and dining areas	47
			Manufacturing or other noisy interior areas	47

TABLE 3-5a (continued)

Type of Occupancy	Room Being Considered	and	Adjacent Area	Sound Isolation Requirement
Normal business offices, drafting areas, banking floors, etc.	Large general office areas		Corridors or lobby	STC-32
			Exterior of building	32
			Data Processing areas	37
			Manufacturing areas and mechanical equipment areas	42
			Kitchen and dining areas	37
Offices in manufacturing, laboratory, or test areas	Shop and laboratory offices		Adjacent offices	STC-37
			Manufacturing, laboratory, or test areas	42
			Washrooms and toilet areas	37
			Corridor or lobby	32
			Exterior of building	32
Motels and urban hotels	Bedrooms		Adjacent bedrooms, separate occupancy	STC-47
			Bathrooms, separate occupancy	47
			Living rooms, separate occupancy	47
			Dining areas	47
			Corridor, lobby, or public spaces	47
			Mechanical equipment rooms	52
			Exterior of building: Normal street or highway noise	42
			Heavy highway traffic	47
			Airport noise	47

TABLE 3-5b Wall, Partition, or Panel Between

Type of Occupancy	Room Being Considered	and	Adjacent Area	Sound Isolation Requirement
Apartments, multiple dwelling buildings	Bedrooms		Adjacent bedrooms, separate occupancy	STC-47
			Bathrooms, separate occupancy	47
			Bathrooms, same occupancy	37
			Living rooms, separate occupancy	47
			Living rooms, same occupancy	42
			Kitchen areas, separate occupancy	47
			Kitchen areas, same occupancy	42
			Mechanical equipment rooms	52
			Corridors, lobby, public spaces	47
			Exterior of building	42
Apartments, multiple-dwelling buildings	Living rooms		Adjacent living rooms, separate occupancy	STC-47
			Bathrooms, separate occupancy	47
			Bathrooms, same occupancy	42
			Kitchen areas, separate occupancy	47
			Kitchen areas, same occupancy	42
			Mechanical equipment rooms	52
			Exterior of building	37

TABLE 3-5b (continued)

Type of Occupancy	Room Being Considered	and	Adjacent Area	Sound Isolation Requirement
Private, single-family residences	Bedrooms		Adjacent bedrooms	STC-37
			Living rooms	42
			Bathrooms, not directly connected with bedroom	42
			Kitchen areas	42
			Exterior of building	37
Private, single-family residences	Living rooms		Adjacent bathrooms	STC-42
			Kitchen areas	42
			Exterior of building	37
Normal school buildings without extraordinary or unusual activities or requirements	Classrooms		Adjacent classrooms	STC-37
			Laboratories	42
			Corridor or public areas	37
			Kitchen and dining areas	42
			Shops	47
			Recreational areas	47
			Music rooms	47
			Mechanical equipment rooms	52
			Toilet areas	42
			Exterior of building	37
Normal school buildings without extraordinary or unusual activities or requirements	Large music or drama areas		Adjacent music or drama rooms	STC-52
			Corridor or public areas	47
			Practice rooms	47
			Shops	47
			Recreational areas	47
			Laboratories	47
			Toilet areas	47
			Mechanical equipment rooms	52
			Exterior of building	47

TABLE 3-5b (continued)

Type of Occupancy	Room Being Considered	and	Adjacent Area	Sound Isolation Requirement
	Music practice rooms		Adjacent practice rooms	STC-47
			Corridors and public areas	47
	Language labora- tories		Same as for theaters, concert halls, auditorium, etc.	
	Counseling offices		Same as for execu- tive offices	
Any occupancy where serious performances are given. Re- quirements may be relaxed for elementary schools or non- critical types of occupancy	Theaters, concert halls, lecture halls		Adjacent similar areas	STC-52
			Corridors and public areas	47
			Recreational areas	52
			Mechanical equip- ment spaces	52
			Classrooms	47
			Laboratories	47
			Shops	52
			Toilet areas	52
			Exterior of building	52
Any occupancy where serious amateur or any professional work is done	Radio, TV, recording studios		Use professional consultants. This is an extremely critical type of area.	

In addition, floors must provide protection against *structure-borne* (impact) sound.

The Sound Pressure Level is measured in the room *below* the test floor on which an impact-producing machine operates. The better the floor, the *lower* the level in the room below.

A "standard contour" which reflects subjective response to noise (approximately the obverse of the STC-contour) is fitted to the test curve; the relative vertical position of the contour determines the "Impact Noise Rating" (INR) of the floor.

The older FHA procedure rated the level shown in Figure 2-10 (Section II, page 41) as "0." Constructions producing levels *below* the standard were rated as "+," indicating superiority over the standard; those producing levels *above* the standard were rated as "–," indicating inferiority. Current procedure adds (algebraically) approximately 51 points to all INR values to convert them to positive numbers identified as "Impact Insulation Class" (IIC).

Current impact test procedures and ratings are somewhat controversial, particularly in the United States. Considerable activity in ASTM committees is directed toward modifying procedures significantly. All data currently available (and probably for some years to come) are determined by tapping machine procedures and calculated from one of the contours shown in this book.

(**Note:** Unless otherwise specifically stated, "floors" in this book refers to the entire "floor/ceiling assembly.")

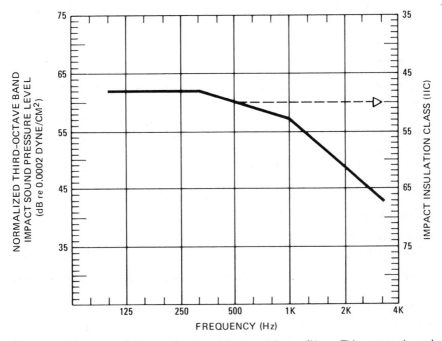

Figure 16-b Determination of Impact Noise Rating of floors. (Note: This contour is used to determine the IIC rating of floor/ceiling assemblies according to ASTM E 492-73T. Note how it differs from Figure 16, Section II, page 41.

Suggested impact isolation criteria for floors immediately *above* the following spaces are given in Table 3-6.

TABLE 3-6 Impact Noise Criteria for Floors

Room Below the Floor	INR RATING	IIC
Multifamily dwelling rooms, hotels, motels, etc.	0	51
Single-family dwelling rooms	No requirement	
School classrooms	−5	46
Offices	−5	46
Auditoriums, lecture rooms, and similar spaces	+5	56
Radio studios, music areas, and similar spaces	Use professional consultants	

(Refer to Section II, page 39, for further discussion.)

SOUND TRANSMISSION LOSS—PARTITIONS AND WALLS

Sound transmission through walls (or floors and ceilings) varies with the *frequency* of the sound and the *weight* (or mass) and *stiffness* of the construction.

Theoretically, the Transmission Loss varies at the rate of 6 dB per doubling (or halving) of the weight of the construction. For example:

Weight (lb/ft^2):		5	10	20	40	80	160
Transmission Loss at 400 cps (Hz) (dB):		33	39	45	51	57	63

In practice, this does not always occur, although it is a reasonable "rule of thumb" for preliminary design.

A single solid panel behaves less well than the simple "mass law" would predict. "Coincidence" effects, related to the panel's stiffness, may degrade the panel's performance appreciably.

A true "double wall" with separate, unconnected wythes performs better than the mass law predicts. The Transmission Loss tends to increase about 5 dB for each doubling of the air space between wythes (minimum air space approximately 2 in [5 cm]). For example:

Air space (in):	3	6	12	24	48
Transmission Loss (dB):	40	45	50	55	60

Resilient attachment of surface "skins" to studs or structural surfaces provides the same effect as separate wythes.

Absorptive material in the cavity between wythes, particularly for lightweight constructions, improves Transmission Loss.

"Viscoelastic" (somewhat resilient but not fully elastic) materials—such as certain insulation boards—"damp" or restrict the vibration of panels. When used with rigid panels (such as gypsum boards), they increase Transmission Loss appreciably. (See Section II, page 58, for a detailed discussion of these subjects.)

The STC-ratings of ordinary "base constructions" for most of the important types of U.S. construction systems are listed in the following tables. The tabulated values represent reasonable averages of many tests by several laboratories and testing agencies and thousands of field tests. While the values may differ slightly from published data from a specific laboratory or manufacturer, they are conservative and representative of what can be expected from a *good* installation of the particular construction described.

For any construction with unique details or characteristics, refer, if possible, to actual laboratory test data and the qualifying descriptive details of the test specimen.

To use these tables, choose the base construction which most closely matches the system which you prefer to use on a particular installation (for example, "dry-wall and wood studs"). Then turn to subsequent pages for typical modifications to the base construction to determine how much improvement in Sound Transmission Loss will be provided by such modifications. Within reasonable limits, almost any STC-rating can be provided by successive improvements to the base construction.

(To convert English units used in the following tables to metric or SI units, the following conversion factors are sufficiently accurate:

1 oz/yd^2	0.034 kg/m^2	1 m	3.3	ft
1 lb/ft^2	5 kg/m^2	1 m^2	11	ft^2
1 lb/ft^3	16 kg/m^3	1 m^3	35	ft^3.)
1 in	2.54 cm			

TABLE 3-7   STC-ratings of Partitions and Walls

2″ × 4″ Wood Studs, 16″ on centers (unless otherwise noted)

Construction	Weight (lb/ft²)	STC-Rating	Plans (3/4″ = 1′0″)
1/4″ Plywood — Nailed to Studs	2-1/2	24	
1/2″ Wood Fiberboard — Nailed to Studs	3-1/2	28	
1/2″ Gypsum Board — Nailed to Studs (Joints Taped and Sealed)	5-3/4	33	

TABLE 3-7 (continued)

Construction	Weight (lb/ft²)	STC-Rating	Plans
3/8″ Gypsum Lath — Nailed to Studs — 1/2″ Sanded Gypsum Plaster (2 Coats)	15	35	
Metal Lath — Nailed to Studs — 7/8″ Sanded Gypsum Plaster (3 Coats)	20	37	

3-5/8″ Metal Studs, 24″ on centers (unless otherwise noted)

Construction	Weight (lb/ft²)	STC-Rating	Plans
5/8″ Gypsum Board — Screw Attachment to Studs (Joints Taped and Sealed)	6	39	

3-1/4″ Metal Studs, 16″ on centers (unless otherwise noted)

Construction	Weight (lb/ft²)	STC-Rating	Plans
3/8″ Gypsum Lath — Clipped to Studs — 1/2″ Sanded Gypsum Plaster (2 Coats)	15	40	
Metal Lath — Clipped to Studs — 3/4″ Sanded Gypsum Plaster (3 Coats)	19	37	

Note: Combinations of channels or similar sections to produce a similar air space between opposite surfaces provide approximately the same STC-ratings.

Arrangements of channels or studs to produce completely independent (nonconnected) wythes provide approximately the same improvement in STC-rating (10 points) as staggered studs.

TABLE 3-7 (continued)

The following modifications to the "base constructions" produce the following improvements in STC-rating:

1. Surface Skin Weight:

Doubling 1 Side. . . . .   +3 points
Doubling 2 Sides. . . .   +5 points

2. Resilient Attachment of Surface Skin:

1 Side. . . . .   +6 points
2 Sides. . . . .   +10 points

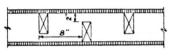

3. Staggered Studs: +10 points

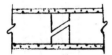

4. Slotted Studs: Not so good now   +8 points

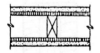

5. Resilient Damping Board
   Layer Under Surface Skins
   (Surface Adhesively Applied):   2 Sides. . . . .   +10 points

6. Absorption in Cavity: +5 points

TABLE 3-7 (continued)

CUMULATIVE IMPROVEMENT OF ANY COMBINATION OF THESE MODIFICA-
TIONS IS CALCULATED THUS:

LARGEST NUMBER + 1/2 NEXT LARGEST + 1/2 NEXT LARGEST, etc.

EXAMPLE:

A "base construction," such as 1/2″ Gypsum Board on Wood Studs (STC-33)

Change to:

Staggered Studs (+10 points)	=	10 points
Add:		+
Absorption in Cavity (+5 points); 1/2(5 points) =		2-1/2 points
Add:		+
One Additional 1/2″ Gypsum Board to One Side Only (+3 points); 1/2(3 points)	=	1-1/2 points
	TOTAL	14 points

Therefore, "base construction" increases from STC-33 to STC-47.

HOLLOW MASONRY BLOCK WALLS

Construction	Weight (lb/ft²)	STC-Rating	Plans
4″ Lightweight[a]	20	36	
4″ Dense	30	38	
6″ Lightweight[a]	28	41	
6″ Dense	43	43	
8″ Lightweight[a]	34	46	
8″ Dense	55	48	
12″ Lightweight[a]	50	51	
12″ Dense	80	53	

[a]Sealed against air leakage with 2 coats of sealer paint both sides or similarly sealed.

TABLE 3-7 (continued)

SOLID MASONRY WALLS

Construction	Weight (lb/ft²)	STC-Rating	Plans
4" Brick[b]	38	41	
8" Brick[b]	80	49	
12" Brick[b]	120	54	
6" Reinforced Dense Concrete	75	46	
8" Reinforced Dense Concrete	95	51	
12" Reinforced Dense Concrete	145	56	

[b]Careful workmanship; airtight joints or surface sealed.

The following modifications to the "base constructions" produce the following improvements in STC-rating:

1. Sand-Filled Cores: +3 points

2. 1/2" Sanded Plaster (or
   Similar Surface Skin):   1 Side.............   +2 points
                                      2 Sides............   +4 points

3. Rigidly Furred Surface Skin: 1 Side.............   +7 points
                                      2 Sides............   +10 points

TABLE 3-7 (continued)

4. Resiliently Attached Surface
    Skin: 1 Side . . . .   +12 points
                                            2 Sides . . . .   +15 points

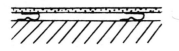

5. Absorption in Cavity: 1 Side . . . .   +3 points
                                            2 Sides . . . .   +5 points

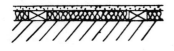

6. Dividing Wall into Separate
    Wythes with 4″ Air Space: +15 points

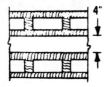

CUMULATIVE IMPROVEMENT OF ANY COMBINATION OF THESE MODIFICA-
TIONS IS CALCULATED THUS:

LARGEST NUMBER + NEXT LARGEST + NEXT LARGEST, etc.

EXAMPLE:

A "base construction," such as 4″ Lightweight Block (STC-36)

    Add:
      Resilient Plaster Skin One Side =   12 points
    Add: +
      Sand in Cores   =   3 points
    Add: +
      Plaster on Opposite Side   =   2 points
                                   TOTAL   17 points

Therefore, "base construction" increases from STC-36 to STC-53.

TABLE 3-7 (continued)

"STUDLESS" CONSTRUCTIONS

Construction	Weight (lb/ft²)	STC-Rating	Plans
2″ Panel, Sanded Gypsum Plaster on Metal Lath With or Without Imbedded Channels	18	34	

Note: Gypsum lath instead of metal lath provides approximately the same STC-rating.

Construction	Weight (lb/ft²)	STC-Rating	Plans
2-1/2″ Panel, Sanded Gypsum Plaster on Separate Layers of Gypsum Lath	19	38	

SPECIAL "DRY-WALL" CONSTRUCTIONS

Construction	Weight (lb/ft²)	STC-Rating	Plans
2-1/4″ Solid Laminated Gypsum Board	10	30	
5/8″ Gypsum Board Layers Laminated to 1-5/8″ × 6″ Gypsum Strips	7	34	
Double Solid Dry-wall — 2 Separate Wythes of 1/2″ Gypsum Board Laminated to 1″ Gypsum Board with 1-1/8″ Air Space	14	46	

TABLE 3-7 (continued)

SOLID, SINGLE-SHEET MATERIALS

Construction	Thickness	Weight (lb/ft²)	STC-Rating
Aluminum	0.025″	.35	19
Plywood	1/4″	.73	22
Cellulose Fiberboard	1/2″	.75	22
Plate Glass	1/4″	3.2	26
Sheet Steel	18 Gauge	2.0	30
Lead	1/16″	3.9	34

Note: To calculate the effect of variations in weight or air space, refer to page 104.

Also, see page 35 for the effect of other construction variables.

CUMULATIVE IMPROVEMENT OF ANY COMBINATION OF THESE MODIFICA-TIONS IS CALCULATED THUS:

LARGEST NUMBER + NEXT LARGEST + NEXT LARGEST, etc.

EXAMPLE:

A "base construction," such as Double Solid Dry-wall (STC-46)

Change to:
Triple Solid Dry-wall

Add:
2 Air Spaces of 1-1/8″   =   5 points
Add: +
1 Layer of 1″ Gypsum Board in Cavity =   2 points
Add: +
Absorption in One Cavity   =   5 points

           TOTAL   12 points

Therefore, "base construction" increases from STC-46 to STC-58.

TABLE 3-7 (continued)

WINDOWS AND GLAZING

Construction	Thickness	Weight (lb/ft²)	STC-Rating
D. S. Glass	1/8″	1-1/2	21
1/4″ Plate Glass	1/4″	3	26
1″ Insulating Glass	1″	6-1/2	32
9/32″ Laminated Acoustical Glass[c]	9/32″	3-1/4	36
Glass Block	3-3/4″	20	40
Spaced Glass (1/4″–2″ Air Space–1/4″)	2-1/2″	6-1/2	42

[c]"Acousta-pane 36"

DOORS[d]

Construction	Thickness	Weight (lb/ft²)	STC-Rating
1-3/4″ Hollow Core Wood	1-3/4″	3-1/2	26
1-3/4″ Solid Core Wood	1-3/4″	5	29
1-3/4″ Hollow Metal	1-3/4″	5	30
1-3/4″ Packed Metal	1-3/4″	7	32
1-3/4″ Special Acoustical	1-3/4″	6	35
2-1/4″ Solid Core Wood	2-1/4″	7	32
2-1/2″ Special Acoustical	2-1/2″	8	38

[d]Fully gasketed, all edges and bottom. "Leaky" gaskets or no gaskets can reduce STC-ratings by 5 to 15 points.

TABLE 3-7 (continued)

MOVABLE AND OPERABLE PARTITIONS

STC-ratings range from STC-18 to STC-48, depending upon construction, weight, and tightness of seals and closures.
Generally, performance parallels comparable fixed-wall construction if edges and perimeters are well-sealed. In practice, it is very difficult to maintain good perimeter seals, and performance tends to be far below ratings.

Floors tend to perform very much like partitions in airborne sound transmission. In impact noise (structure-borne) transmission, their response is similar but not identical; a floor with a good STC-rating may not have a good INR-rating. Therefore, each characteristic must be considered independently.

TABLE 3-8 STC-ratings and INR-ratings of Floors

2″ × 10″ Wood Joists, 16″ on centers (unless otherwise noted)

Construction	Weight (lb/ft²)	STC-Rating	INR-Rating	Plans
1/2″ Plywood Subfloors and Standard Oak Flooring — Nailed to Joists	8	25	−28	
Same, Plus 5/8″ Gypsum Board Ceiling — Nailed to Underside of Joists	10	37 (+12)	−17 (+11)	
Same, Except 3/8″ Gypsum Lath and 1/2″ Sanded Plaster	15	39	−15	
Same, Except Metal Lath and 7/8″ Sanded Gypsum Plaster (3 Coats)	17	39	−15	

TABLE 3-8 (continued)

The following modifications to the "base constructions" produce the following improvements in STC- and INR-ratings:

	STC	INR
1. Resilient Suspension of Ceiling:	+10	+8
2. "Floating Raft" — Rough Flooring and Finish Flooring on 1″ × 3″ Sleepers Resting On but Not Nailed Through Resilient Fiber Board:	+10	+8
3. Staggered Joists — Ceiling Independent of Floor:	+8	+7
4. Resilient Damping Board Layer Between Subfloor and Finish Floor Underlayment (Underlayment Adhesively Applied to Damping Board):	+7	0 to +2

TABLE 3-8 (continued)

	STC	INR
5. Absorption in Cavity: ([a]only when ceiling resiliently suspended or on staggered joists; little or no improve- ment in rigid constructions)	+3	+7[a]

	STC	INR
6. Vinyl Tile:	0	0
3/32″ Linoleum:	0	+5
1/4″ Cork Tile:	0	+10 to +15
Carpet and Pad:	0	+20 to +25

CUMULATIVE IMPROVEMENT OF ANY COMBINATION OF THESE MODIFICA-
TIONS IS CALCULATED THUS:

LARGEST NUMBER + NEXT LARGEST + NEXT LARGEST, etc.

EXAMPLE:

A "base construction," such as 2″ × 10″ Joists, with 1/2″ Plywood Subfloors
and Standard Oak Flooring with 5/8″ Gypsum Board Ceiling nailed to joists
(STC = 37; INR = −17)

	STC	INR
Add: Resilient Suspension of Ceiling =	10 points	8 points
Add:	+	+
Heavy Carpet on Thick Pad =	0	25 points
TOTAL	10 points	33 points

Therefore, "base construction" increases from STC = 37; INR = −17 to
STC = 47; INR = +16.

TABLE 3-8 (continued)

SOLID REINFORCED CONCRETE SLABS

Construction	Weight (lb/ft²)	STC-Rating	INR-Rating	Plans
4" Reinforced Dense Concrete	50	41	—17	
6" Reinforced Dense Concrete	75	46	—17	
8" Reinforced Dense Concrete	95	51	—16	

RIB3ED CONCRETE FLOORS

Construction	Weight (lb/ft²)	STC-Rating	INR-Rating	Plans
Ribbed Concrete — 2" Slab on 4" Hollow Filler Blocks	65	45	—22	
Ribbed Concrete — 2" Slab on 6" Hollow Filler Blocks	80	49	—21	
Ribbed Concrete — 2-1/2" Slab on 6" Ribs 24" On Centers	55	41	—17	
Ribbed Concrete — 4" Slab on 6" Ribs 24" On Centers	75	46	—17	

TABLE 3-8 (continued)

CONCRETE ON BAR JOISTS[b]

Construction	Weight (lb/ft²)	STC-Rating	INR-Rating	Plans
2″ to 2-1/2″ Concrete on Lath or Light Metal Forms	35	37	—24	

CONCRETE ON CELLULAR METAL FLOORS[b]

Construction	Weight (lb/ft²)	STC-Rating	INR-Rating	Plans
2″ Concrete on Light Cellular Metal Floor	35	38	—22	

[b]These floors tend to be particularly susceptible to horizontal "flanking" unless special precautions are taken at edges and supports. (See page 127.)

The following modifications to the "base constructions" produce the following improvements in STC- and INR-ratings:

	STC	INR
1. "Floating Raft" — Rough Flooring and Finish Flooring on Sleepers Resting on Resilient Fiber Board or Blanket or on Rubber or Spring Clips:	+12	+20
2. Same — but Sleepers Directly on Concrete; No Resilient Material:	+7	+15

TABLE 3-8 (continued)

3. "Floating" Concrete Topping on 1" Thick Glass Fiber Mat or Equivalent: +10   +15

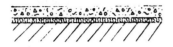

4. Suspended Ceiling on Resilient Runners, Resilient Clips, or Wire Hangers; Substantial Air Space: +12   +8

5. Rigidly Furred Ceiling Skin; Very Small Air Space: +7   0

6. 1/2" Sanded Plaster or Similar Surface Skin: +2   0

7. Resilient Damping Board Under Flooring: 0   +10

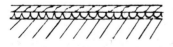

8. Absorption in Cavity (only when ceiling resiliently suspended; little or no improvement in rigid constructions):   +3   +5^c

9. Wood Finish Flooring at Least 1/2" Thick; Set in Mastic: 0   +7

TABLE 3-8 (continued)

10. Vinyl Tile:[d]	0	0
3/32" Linoleum:	0	+3 to +5
1/4" Cork Tile:	0	+10 to +15
Carpet and Pad:[d]	0	+20 to +30

([d]except on concrete on bar joists or concrete on cellular metal floors, where values are:

Vinyl Tile:	0	+5
Carpet and Pad:	0	+25 to +40)

CUMULATIVE IMPROVEMENT OF ANY COMBINATION OF THESE MODIFICATIONS IS CALCULATED THUS:

LARGEST NUMBER + NEXT LARGEST + NEXT LARGEST, etc.

EXAMPLE:

A "base construction," such as 6" Reinforced Dense Concrete, Bare Surface (STC = 46; INR = −17)

	STC	INR
Add:		
"Floating Raft" =	12 points	20 points
Add:	+	+
3/32" Linoleum =	0	4 points
TOTAL	12 points	24 points

Therefore, "base construction" increases from STC = 46; INR = −17 to STC = 58; INR = +7.

Partial Height Barriers

When complete enclosure is impossible or impractical, partial height barriers or "screens" may afford some protection if properly used. Usually, less than 20 dB protection, and then only in the higher frequencies, is the maximum attainable with this procedure. (Refer to Section II, page 44, for a discussion of partial height barriers and to Section III pages 147 and 192.)

Special High Isolation Constructions

For highly critical spaces (radio, TV, and recording studios, testing laboratories, etc.), complex, high-performance constructions are used to isolate against airborne and structure-borne noise and vibration. "Floating" floors (and even floating rooms), double-wythe constructions, and complex and carefully detailed systems are usually employed for such work. It is always advisable to use the services of a highly qualified, experienced consultant for such work; it should

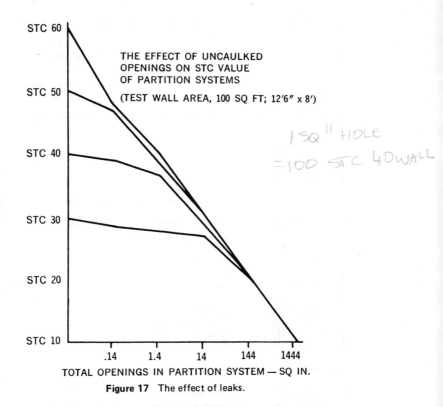

Figure 17 The effect of leaks.

never be attempted by the inexperienced, since the construction is very expensive and minor mistakes or oversights can literally ruin a job.

"Leaks and Flanking"

All of the performance data in the preceding tables assume tight, careful construction. However, this assumption is rarely justified in practice. Even very good constructions can be seriously degraded by the traditional minor carelessnesses and overlooked details all too typical of the building and industry (see Figure 17). *We cannot overemphasize the importance of this subject!* The following pages discuss typical problems, how to avoid them, and how to correct them. The reader is urged to read carefully the discussion of these matters in Section II, pages 41-44.

One square inch (6.5 cm^2) of hole through an STC-40 wall will transmit as much acoustical energy as almost *100 square feet* (9 m^2) of the wall! Shrinkage cracks, openings around any wall penetration, even the "minor" joints and holes are serious! Avoid them! Caulk everything!

As Figure 3-6 indicates, the first four caulking beads produce an enormous improvement in the partition performance; thereafter, there is a rapidly diminishing return.

edge detail		lab STC performance on experimental perimeter
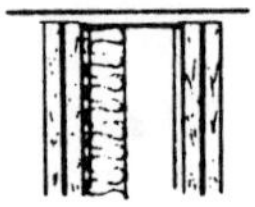	Unsealed	19
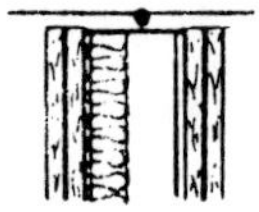	Single bead of caulking under track	30
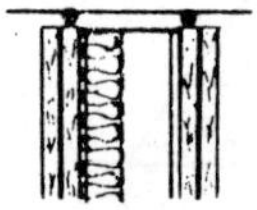	Single bead of caulking under inner layer of gypsum board	48
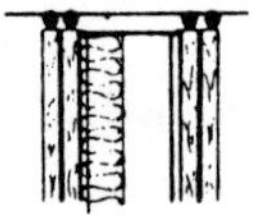	Two beads of caulk each side	51
	Two beads of caulk each side and one under track	52
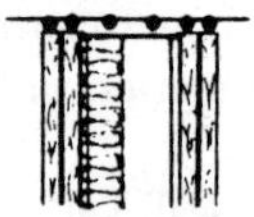	Six rows of caulk bead, two at each side and two under track	54

Figure 3-6   The effect of caulking. (Courtesy United States Gypsum Company.)

See following details for recommended caulking of common acoustical "leaks" often found in usual construction practice.

Remember, if air can get through, sound will!

Figures 3-7 and 3-8 show typical "leaks" and how to prevent them.

Door Gaskets and Perimeter Seals

The performance of movable or operable sound barriers is so strongly dependent upon the gaskets and perimeter seals that it is almost pointless to discuss STC-ratings or Sound Transmission Loss of such barriers *except in their normal operable condition.*

Usually compression seals or gaskets require considerable compression to make them work as sound seals; and it is questionable whether the force necessary to keep them properly compressed is practical (for example, personnel doors which must be "slammed" or pushed shut with difficulty will inevitably be "adjusted" to close easily, ruining the acoustical seal). Movable or operable partitions with adequate sweep seals may operate with such difficulty that they will be unacceptable to the user. Drop seals (automatic threshold closures) often become warped or bent, making them useless.

For large, operable walls, mechanically compressed (or hydraulically operated) seals are usually the most satisfactory. For small personnel doors, a combination of multiple sweep seals and very soft compression seals is usually satisfactory.

Where perimeter seals are important, it is imperative that the floors be very level and regular, and that frames be square, plumb, and straight. Warped wood doors or sprung metal doors rarely provide acceptable acoustical seals.

Poor or even "typical" perimeter seals may degrade the barrier performance by *5 to 15* points!

Avoid interconnected openings, voids, or chases; break or seal off common spaces or constructions. Blocking and discontinuities at the perimeter of each critical space minimize structural flanking (see Figure 3-9).

Continuous Open Plenums

Caution: It is important to recognize that most acoustical absorbents are very poor sound barriers. They are usually porous and lightweight—quite the reverse of what is required to reflect or isolate sound.

A potentially serious leak often occurs at the joint between the ceiling and the top of the partition (Figure 3-10); and the "over-the-top" path (when partitions extend only to or just barely through the ceiling) may be poor. This path should provide at least as much attenuation as the wall or partition if a balanced construction is to be achieved.

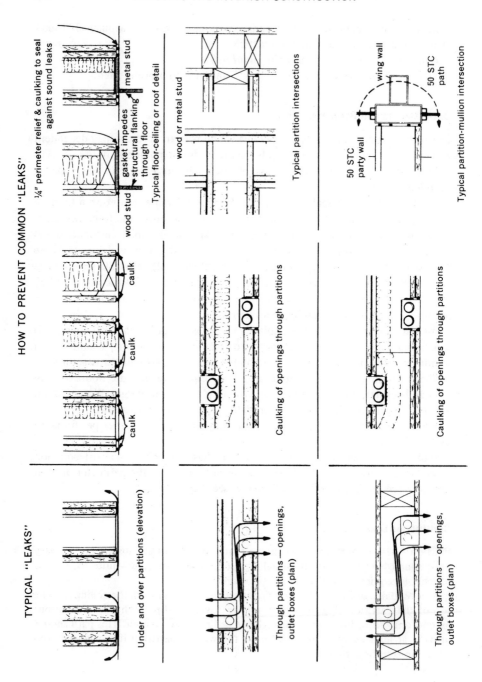

HOW TO PREVENT COMMON "LEAKS"

TYPICAL "LEAKS"

¼" perimeter relief & caulking to seal against sound leaks

metal stud

gasket impedes structural flanking through floor

wood stud

Typical floor-ceiling or roof detail

wood or metal stud

Typical partition intersections

wing wall

50 STC path

50 STC party wall

Typical partition-mullion intersection

caulk

caulk

caulk

Caulking of openings through partitions

Caulking of openings through partitions

Under and over partitions (elevation)

Through partitions — openings, outlet boxes (plan)

Through partitions — openings, outlet boxes (plan)

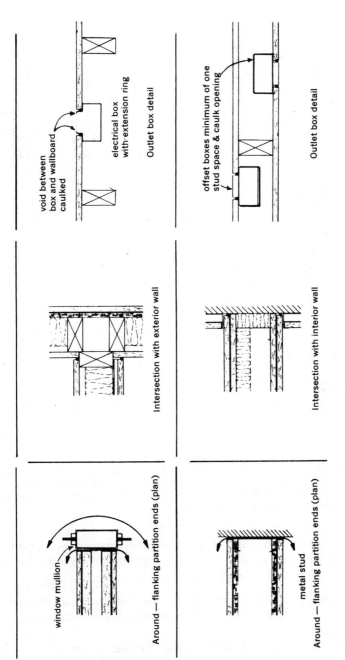

Figure 3-7 Caulking details. (Courtesy United States Gypsum Company.)

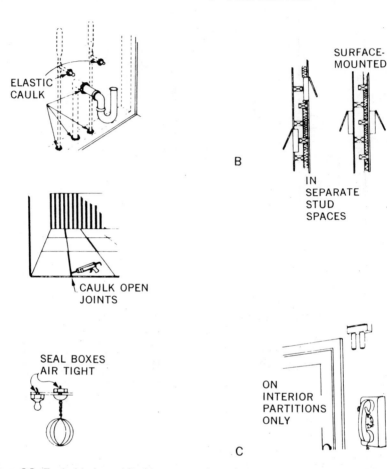

Figure 3-8 Typical leaks and flanking sources. A, typical potential "leaks"; B, typical penetrations: Surface-mount cabinets, or put them into separate stud spaces. Caulk all perimeters! C, typical wall-mounted "noise-makers": Mount such noise sources only on interior partitions or noncritical walls or other surfaces.

When the plenum is to be used as a continuous return air plenum or when it is necessary to use easily removed partitions which cannot penetrate the ceiling, several measures to improve isolation are available:

1. The ceiling transmission loss may be improved by coating the upper side of the tile; or a septum or barrier of some type may be laid over the upper side of the tile.

2. A thick (at least 3 in [8 cm]) insulation blanket or batt layer may be laid over the entire ceiling. This normally provides 5 to 8 dB additional attenuation. A strip of this same insulation, laid over the top of the ceiling but extending

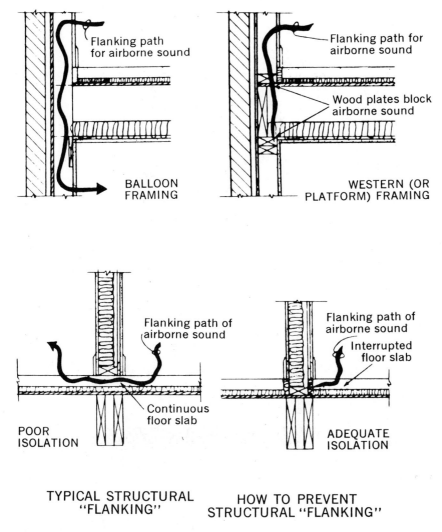

POOR

GOOD

Flanking path for airborne sound

Flanking path for airborne sound

Wood plates block airborne sound

BALLOON FRAMING

WESTERN (OR PLATFORM) FRAMING

Flanking path of airborne sound

Flanking path of airborne sound

Interrupted floor slab

Continuous floor slab

POOR ISOLATION

ADEQUATE ISOLATION

TYPICAL STRUCTURAL "FLANKING"

HOW TO PREVENT STRUCTURAL "FLANKING"

Figure 3-9 Typical flanking paths. (Courtesy United States Gypsum Company.)

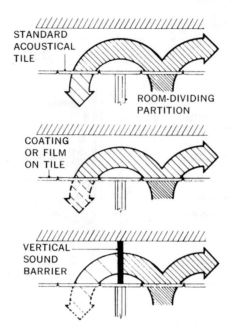

STANDARD
ACOUSTICAL
TILE

ROOM-DIVIDING
PARTITION

COATING
OR FILM
ON TILE

VERTICAL
SOUND
BARRIER

Figure 3-10   Flanking via the ceiling.

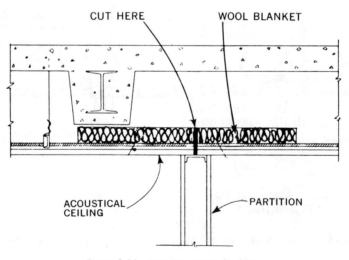

CUT HERE   WOOL BLANKET

ACOUSTICAL
CEILING

PARTITION

Figure 3-11   Avoiding ceiling flanking.

only 48 in (1.2 m) on *both* sides of the partition, frequently provides enough additional attenuation (3 to 5 dB) to minimize the problem in existing installations where other measures are not practical (Figure 3-11).

3. Erecting a vertical septum or barrier from the top of the partition to the slab above is very effective, if it does not interfere with return air flow. If air flow is a problem, various lined or muffled transfer duct arrangements through the septum may be used. If the ceiling may be pentrated at all, carrying only one surface of the partition to the slab above (rather than carrying the complete partition construction all the way up) is normally an adequate means of providing a "septum." Even penetrating the ceiling enough to carry the partition at least 6 in (15 cm) above the ceiling, and then laying insulation over the ceiling as described above, significantly improves isolation.

4. If a lightweight ceiling (even single-layer gypsum board) or the metal supporting runners of an acoustical ceiling are continuous over the top of the partition, a shear wave, capable of transmitting energy to adjacent spaces, often develops in the ceiling diaphragm. This usually limits the ceiling attenuation to STC-38 or much less. Cuting through the gypsum board or the runners directly over the top of the partition often improves the ceiling attenuation by as much as 5 to 10 dB (Figure 3-11). (Refer to the current annual *Sweet's Catalog* or manufacturers' catalogs for ceiling attenuation factors—equivalent to STC rating for barriers.)

The "Weak Link"

A door, window, or similar insertion into a wall or barrier usually represents a "weak link" in the construction, since rarely will its performance match that of the wall construction. A good "rule of thumb" for designing a "balanced" construction (wall plus any tight, well-caulked, or gasketed fenestration or insert) is:

1. If the area of the insert is *less than 25%* of the total wall area in which it is inserted, its STC-rating may be up to 5 points lower than the STC of the wall without seriously degrading the barrier's performance.
2. If more than 25% and *less than 50%*, its rating may be 2 points lower than that of the wall.
3. If *more than 50%*, its STC-rating will essentially determine the performance of the entire barrier.

(Refer to Section II, page 43, for further discussion of this subject.)

Ordinary glazing, hollow-core doors, plywood panels, and similar thin, lightweight panels are especially poor barriers; they invariably are the weak link in a wall. (See Table 3-7, page 113.)

Generally, small stiff panes of glass show significantly higher STC ratings than large, more flexible panes. Elastomeric edge gaskets (such as "zipper" types) with a good grip (at least $\frac{5}{8}$ in. [1.6 cm]) on the edge of the glass can increase the STC performance of a window by as much as 4 dB.

Privacy

"Privacy" is largely a subjective thing, meaning something different to almost everyone. Typical "high-privacy" spaces include offices for attorneys, doctors, psychologists, counselors, financial officers, etc. European mores apparently dictate higher privacy for "private" offices than American custom with its "open-door policy" even in executive spaces.

In acoustical design, privacy usually refers to sufficient attenuation of a signal to prevent intelligibility of speech (whether the conversation must be kept *in* or *out*).

Most of the speech intelligibility is carried in the 300 to 5000 cps (Hz) frequency range; actual speech "privacy" is largely determined by the attenuation provided in this range.

A reasonable design "rule of thumb" for determining privacy is to add the STC of the enclosing barriers (walls, floor/ceiling assembly, etc.) to the background (or masking) level in dBA. The degree of expected privacy falls into these ranges:

High	STC + dBA (background) = 85 or higher
Medium	STC + dBA (background) = 80
Low	STC + dBA (background) = 75 or lower

Masking

We tend to hear by "contrast"—that is, when a signal is sufficiently louder than the background sound to be intelligible. The background level in any space must be low enough to permit the occupants to hear wanted sound and to communicate. (See "Background Noise Criteria," Section III, page 88.) Intruding noise, as well as background sound within spaces, must be low enough in level not to interfere with tasks within the space.

An objective measure of speech interference is the "Speech Interference Level"—the arithmetic average of the Sound Pressure Level in the octaves centering on 500, 1000, and 2000 cps (Hz) (or the four octaves from 300 to 4800 cps (Hz]). When this level becomes excessive (in excess of 65 dB), speech communication within a space becomes difficult.

However, some noncritical spaces are too quiet! Even low-level sounds intrude objectionably into them. When it is impossible or impractical to provide enough isolation between spaces with a barrier, it is occasionally helpful to raise the background level within a space to where intruding sound from adjacent areas is no longer intelligible—that is, it is "masked."

This may be done by adding "masking" sound to the space, in the form of steady, broadband, neutral sound, such as air diffuser noise, splashing fountains, etc. Music is rarely effective, since it contains many pauses and relatively pure, discrete tones.

Generally, an acceptable masking level must be less than 47 dBA, and the masking sound spectrum must be "tailored" to the voices or sounds which are to be masked.

Masking is a useful, but complex and limited, procedure; normally it should be attempted only by experienced professionals, lest it cause more problems than it solves. It must not be used in spaces such as radio and TV studios, concert halls, etc., where low background levels are imperative.

(Note: Occasionally, relief from intruding noise can be provided by better furniture arrangement in adjacent spaces. Take the TV sets off the party wall, and don't arrange desks so that occupants sit back-to-back on the party wall, etc.)

SOUND ABSORPTION

Sound absorbents are used:

1. To reduce the Sound Pressure Level within spaces.
2. To prevent reflections from surfaces.
3. To control reverberation within spaces.

In general, absorbents should be located:

1. As near the offending sound source as possible.
2. On surfaces producing unwanted reflections.
3. On surfaces not required for helpful reflections.

Sound-absorbent materials include the "*natural*" absorbents (carpets, draperies and fabrics, upholstered furniture, people, etc.) and the *applied* absorbents (acoustical tiles, panels, blankets, etc.). Rough textures, thin "fuzzy" or fibrous surfaces, special paints, and similar items are *not* effective sound absorbents.

The following tables provide a reasonable guide to the selection and location of acoustical absorption for typical spaces. For critical or special areas, always obtain the recommendations of a competent professional.

Remember:

1. The ceiling may *not* be the proper location for absorption.
2. Many spaces, particularly acoustically critical spaces, often require *no* applied absorption. Natural absorption provided by seats, furnishings, the audience, etc., may be adequate; or special materials may be required to supplement the natural absorbents in certain frequency ranges.
3. Some spaces require variable absorption (such as multipurpose spaces for various musical presentations as well as for drama). Drapes, on a traverse, are typical of such absorption.

TABLE 7   Common Porous Absorbents[a]

Material	Thickness (in)	Thickness (cm)	Density (lb/ft³)	Density (kg/m³)	Noise Reduction Coefficient
Mineral or glass wool blankets	$\frac{1}{2}$ to 4	1.3 to 10	$\frac{1}{2}$ to 6	8 to 100	.45 to .95
Molded or felted tiles, panels, and boards	$\frac{1}{2}$ to $1\frac{1}{8}$	1.3 to 3	8 to 25	125 to 400	.45 to .90
Plasters	$\frac{3}{8}$ to $\frac{3}{4}$	1 to 2	20 to 30	320 to 480	.25 to .40
Sprayed-on fibers and binders	$\frac{3}{8}$ to $1\frac{1}{8}$	1 to 3	15 to 30	240 to 480	.25 to .75
Foamed, open-cell plastics, elastomers, etc.	$\frac{1}{2}$ to 2	1.3 to 5	1 to 3	16 to 48	.35 to .90
Carpets	Varies with weave, texture, backing, pad, etc.				.30 to .60
Draperies	Varies with weave, texture, weight, fullness, etc.				.10 to .60

| | Absorption Coefficients of Floor Area at: | | | | | |
	125	250	500	1000	2000	4000 cps (Hz)
Seated audience	.60	.75	.85	.95	.95	.85
Unoccupied upholstered (fabric) seats	.50	.65	.80	.90	.80	.70

[a](There are literally hundreds of absorbents available today; their performance varies widely and changes constantly. For current data, refer to the annual *Sweet's Catalog*.)

(Refer to Section II page 45, for a detailed discussion of absorbents and their functions and performance characteristics.)

Amount, Type, and Location of Absorption

The following tables are conservative "rule-of-thumb" recommendations for the use of absorption in common occupancies. For further information, refer to various books and publications, or to a competent acoustical consultant.

Sound Absorption Coefficient

The ratio of acoustic energy absorbed to the energy incident upon a surface. Usually measured at 125, 250, 500, 1000, 2000, and 4000 cps (Hz), and reported as a decimal or percent (e.g., .65 or 65%).

Noise Reduction Coefficient

The arithmetic average of the sound absorption coefficients at 250, 500, 1000, and 2000 cps (Hz).

TABLE 3-9 Amount, Type, and Location of Absorption

| Room | Ceiling[a] | | | | | Wall Treatment[b] | Special[c] |
| | Full | Partial | NRC Range | | | | |
			.45 to .65	.65 to .75	over .75		
Private offices	X			X			
General office space	X			X			
Computer and accounting rooms	X				X	X	
Classrooms, elementary	X			X			
Classrooms, secondary and college		X		X		X	
Language laboratories	X			X		X	X
Libraries	X			X			
Laboratories	X			X			
Meeting and conference rooms		X		X		X	
Gymnasiums, arenas, and recreational spaces	X			X			X
School and industrial shops, factories, etc.	X				X	X	X
Stores and commercial shops	X			X			
Kitchens	X				X		

TABLE 3-9 (continued)

Room	Ceiling[a]					Wall Treatment[b]	Special[c]
			NRC Range				
	Full	Partial	.45 to .65	.65 to .75	over .75		
Restaurants	X			X		X	
Corridors	X			X		X	
Lobbies	X			X			
Residential living rooms	X		X				
Residential bedrooms	X		X				
Hospital rooms	X			X			
Churches							X
Auditoriums							X
Concert halls							X
Theaters							X
Lecture rooms							X
Mechanical equipment rooms							X
Radio, recording, and T. V. studios							X

[a]Or an equivalent amount of distributed absorption provided by carpets, draperies, furnishings, etc.

[b]Sidewall treatment advisable in addition to ceiling treatment to reduce reflections, flutter, or echo, to further reduce noise, or to control reverberation.

[c]Highly complex applications, usually requiring services of an acoustical consultant. May require special forms or types of absorption, or, in some instances, none at all.

Noise Reduction Calculations

The effect of introducing absorption into a space can be calculated with reasonable accuracy as follows:

$$NR = 10 \log_{10} \frac{A_o + A_a}{A_o}$$

where

A_o = original absorption present in sabins;
A_a = added absorption in sabins;
NR = Sound Pressure Level *reduction* in dB.

(**Note:** The surface area multiplied by the absorption coefficient = the sabins of absorption. When added absorption covers an existing surface, the coefficient of the added absorption must be reduced by the coefficient of the existing surface covered by the added absorption. In a highly absorptive space, the effect of added absorption is small, compared with its effect in a nonabsorptive space. In practice, less than 10 dB Noise Reduction can be obtained by the use of absorption alone.)

Reverberation Time

The reverberation time within a space depends upon the amount of absorption present in the space. A chart of so-called "optimum" reverberation time for rooms, and the well-known "Sabine formula" for calculating reverberation time follow.

"Reverberation Time" means the time for the source signal energy level to decay to one-millionth of the original level (that is, to decay 60 dB). It can be measured or it can be calculated from the Sabine formula:

$$T = 0.05 \frac{V}{A}$$

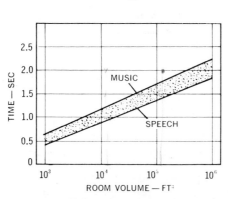

Figure 3-12 Optimum reverberation time.

where

T = time in seconds;
V = volume in cubic feet;
A = total absorption present in sabins (area in square feet of all materials present multiplied by their respective absorption coefficients);

or:

$$T = 0.161 \frac{V}{A}$$

where

T = time in seconds;
V = volume in cubic meters;
A = total absorption in metric sabins within the space.

Normally, smaller rooms should have shorter times than larger rooms; and music spaces usually require longer times than spaces used principally for speech. (Refer to Section II, page 57, for further discussion of reverberation and reverberation time.)

This is one of the overemphasized parameters of acoustical design. In a normal space, designed properly for other acoustical requirements (control of echoes, reduction of noise, etc.), acceptable reverberation time results almost automatically.

In an auditorium-type space with upholstered seats (or with all seats occupied) and an absorptive rear wall (to prevent echo), reverberation time will be about:

$$T = \frac{H}{20}$$

where

T = reverberation time in seconds;
H = average ceiling height in feet.

or:

$$T = \frac{H}{6}$$

where

T = reverberation time in seconds;
H = average ceiling height in meters.

Other requirements (sight lines, lighting, projection, etc.) will normally determine minimum ceiling height, so reverberation time of the occupied space can be approximated from this formula. Usually, for music areas, it will be found that the ceiling must be raised higher than sight lines would require if adequate reverberation is to be provided.

Many spaces require variable reverberation, and it is wise to provide movable draperies or other provisions for changing the absorption present in the spaces.

For spaces where accurate determination and control of reverberation time are important, it is wise to use the services of a qualified acoustical consultant. Such spaces are probably critical from many other acoustical aspects, also; only the experienced practitioner should attempt their design.

SHAPE AND CONFIGURATION

All of the surfaces which enclose a space affect the acoustics within the space. Surfaces absorb, reflect, focus, diffuse, or diffract the sound which reaches them. For most purposes, the path of a sound wave front can be thought of as similar to a ray of light, and optical analogies can be used to study the behavior of sound within an enclosure. (While this is not strictly correct, it is sufficiently accurate to be useful in acoustical design.)

Generally, pure geometric shapes tend to be troublesome. Particularly dangerous are spherical, ellipsoidal, cubical, cylindrical, and similar configurations. Unfortunately, architectural design tends strongly toward such shapes. A discussion of some of the more common problems associated with shape and configuration follows.

Size

Small spaces normally present few difficult problems, but the severity of acoustical problems tends to increase with the size of the room. In small rooms, time of travel of the wave front from source-to-listener, source-to-surfaces-to-listener, and surface-to-surface is short; distinct echoes rarely occur, and reverberation time is usually quite short. Although sound in them may be far from optimum, small spaces are rarely unusable because of their internal acoustics. (Intruding noise, however, can be almost as serious in small spaces as in large spaces.)

In large spaces, echoes, flutter, excessive reverberation, nonuniform distribution of energy, inadequate levels in areas remote from the source, excessive concentration or focusing in some areas ("hot spots"), and similar faults are very common.

Proportion

For optimum response to a broad range of sources, a room should have dimensional proportions approaching $1:\sqrt{2}:\sqrt{3}$. Considerable variation from these proportions is permissible, but it is obvious that many "favorite" shapes cannot be acoustically good for most spaces.

Diffusion

Usually, a diffuse sound field is preferable to a highly specular field (comparable with lighting requirements in a room where soft, diffuse surfaces are usually preferable to shiny surfaces). The relatively bare, "stark," simple surfaces of modern architectural design tend to cause acoustical problems. As much texture as possible is usually desirable in large reflective surfaces.

Diffraction

Small, uniform, regularly spaced reflective surfaces can produce objectionable diffraction effects (similar to optical diffraction gratings). Particularly troublesome are small, regularly spaced slats; small, regularly spaced "clouds" (smaller than 48 in. X 96 in. [1.2 m X 2.4 m]); and regular rows of hard seats or seating risers. Either "tuned" reflections or appreciable selective (tuned) absorption may result when such designs are used.

Reflections

Normally, reflections are undesirable only when they arrive late; if they arrive soon after the direct sound from the source, they reinforce and enrich the sound.

Sound in air travels about one foot in <0.001 sec. A distinct echo is heard if reflected sound follows direct sound by more than 0.060 sec. Thus, if the path difference between direct and reflected sound exceeds 70 ft (21 m) distinct echoes will occur.

Reflections which arrive within 0.020 sec of the direct sound (path difference of about 22 ft [7 m]) are normally highly desirable. In concert halls and spaces used for musical performances by large groups of musicians, reflections arriving less than 0.040 sec after direct sound are usually acceptable and often desirable.

(**Note:** In test spaces and "anechoic" chambers, reflections of any kind are normally undesirable.)

The "sending end" of any space and the surfaces near the source should normally be reflective, and arranged so that the sound reflecting from them reaches listeners soon after the direct sound. Even the central portion of the ceiling in a large classroom or conference room often should be hard and reflective rather than absorptive.

"Band shells," stage enclosures, and "flying" reflectors in stage houses are particularly useful means of reinforcing the direct sound of the performers.

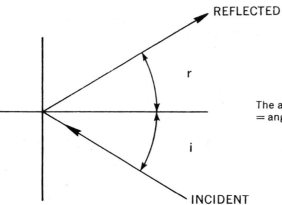

REFLECTED

r

i

INCIDENT

The angle of incidence (i)
= angle of reflection (r)

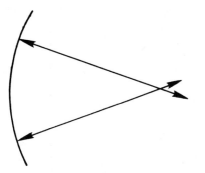

Concave surfaces tend to
focus sound.
(Usually poor in most rooms)

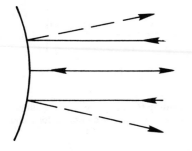

Convex surfaces tend to
diffuse sound.
(Usually good in most rooms)

Figure 3-13 Reflection of sound.

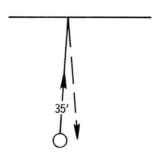

Figure 3-14 Echo. Distinct "echoes" may result when reflective surfaces are 35 ft (10m) or more away from source, and facing source.

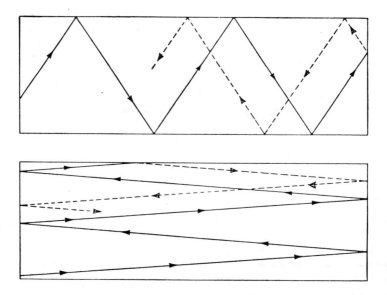

Figure 3-15 Flutter: the "rattle" or "buzz" sound resulting from discrete, rapid, multiple echoes from closely spaced, parallel, reflective surfaces; common in corridors, long and narrow rooms, and similar spaces.

Good, low-cost, portable and adjustable units, as well as large, expensive permanent constructions, are available for this purpose. Pulpit canopies, too, are useful in church design. They must be built of heavy, stiff panels, however; light, flexible reflectors may become selective absorbers (normally very undesirable). (See Figure 3-13.)

(Refer to Section II, page 29, for further discussion of reflection, diffusion, and focus of sound.)

(Refer to Section II, page 27, for further discussion of these problems.)

The Acoustical Design of Specific Spaces

We have assiduously avoided any suggestion that there is *a* design for *a* particular room. That would imply that the design comes from the name rather than the function of the space. A "school," for example, can mean almost anything; it contains spaces to:

> Eat
>
> Play
>
> Listen
>
> Discuss
>
> Sing and perform
>
> Work with tools, office machines, etc.,
>   and many other activities

The same thing is true of a "home," a church, or almost any structure which man builds. There is an optimum acoustical environment for almost any space; the name of the institution does not determine that environment—the activity within the space does.

The "open plan" school, for example, is a very different institution from the conventional school—not just in construction, but in function, concept, philosophy, and program. It is *not* merely a conventional school building with the partitions knocked down. A school for the deaf or blind is quite different from that for normal children.

The principles which govern the acoustical design of any space evolve from the function of that particular space (that is why the "multipurpose" room is usually a disappointing compromise; it serves no purpose very well). The acoustical requirements of spaces include, in varying degrees:

> Quiet
>
> Privacy

Good listening

Adequate loudness of signal

Uniformity of distribution of sound

Fidelity and realism of sound

Intelligibility of speech

Sound control consists not in "sticking on some treatment" or in making spaces "look acoustical," but in:

Providing wanted sound

Eliminating unwanted sound

Eliminating noise at the source

Controlling reverberation

Reducing impact noise and excessive
vibration

Eliminating excessive internal noise

Eliminating intruding, extraneous noise

Controlling signal level, reinforcing
the signal, when necessary

Conserving the hearing of occupants

Obviously, these principles apply to any space, with varying degrees of emphasis. Therefore, there is no optimum design for any *type* of space; there are optimum designs for *each* space.

Where possible, general design principles are indicated in the preceding charts and tables. General criteria are given for the usual types of spaces; and widely applicable design principles are discussed. Performance requirements for materials and constructions are shown, wherever possible.

The following pages illustrate and discuss shaping and "treating" the surfaces of some specific types of spaces, since the principles are broadly applicable to any spaces with similar functional requirements. However, until a designer is reasonably experienced in designing such spaces, he may find it wise to get assistance from an experienced acoustical consultant.

Techniques, materials, and performance data change constantly; tables listing such information tend to be obsolete before a book is off the press. It is imperative that current, up-to-date data be obtained from the best possible sources. But principles do not change; it is important that the professional know and understand them.

Shaping Special Spaces

In general, let reflective surfaces face absorptive surfaces. Distributed absorption is usually preferable to only ceiling absorption. Some reflective surfaces are usually desirable.

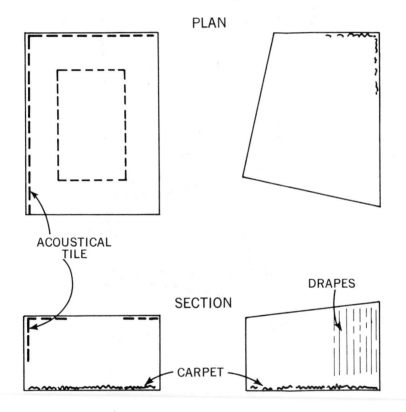

Figure 3-16 Left: Conference, seminar, and classrooms are usually rectangular. Reflective surfaces can provide useful reflections. Right: Studios, music practice rooms, and similar areas may be made with no surfaces parallel. Surfaces may be reflective, absorptive, or adjustable.

Shaping Music Rehearsal Spaces

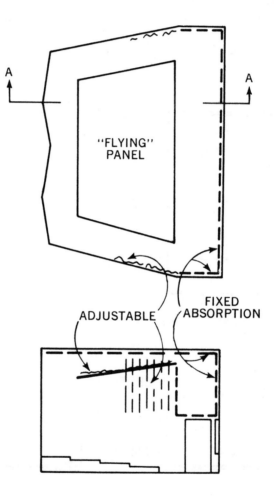

Figure 3-17 Upper, plan: Hard, reflective, splayed rear and sidewalls; reflective "flying" panel over high-pitched instruments; soft, absorptive front walls and front sidewalls; entire ceiling absorptive; adjustable absorption (draperies) optional for splayed sidewalls; absorption (mineral or glass wool blanket) optional above flying panel. Lower, section: Flying panel sloped up toward front of room; ceiling height not less than 14 ft (4.3 m)—16 ft (5 m) better.

Shaping Auditorium-type Spaces

Generally, such spaces should be roughly "egg-shaped," with the "sending end" at the point and the "receiving end" at the blunt end. Avoid broad, shallow rooms or wide fan-shaped plans.

The sending end should be hard and reflective, and the source should "look into" absorptive areas—audience and soft rear wall. Usually sidewalls and ceiling planes should be hard and reflective.

Every seat should have a clear view of the source and an unobstructed view of the sound system speaker.

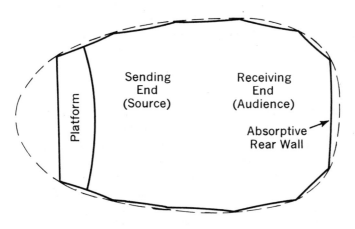

PLAN

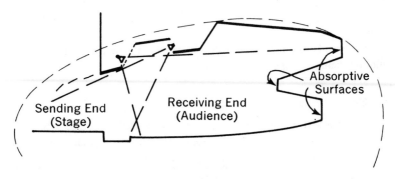

SECTION

Figure 3-18

Additional Requirements

Good sight lines, provided by sufficiently steeply raked seating and a raised platform or stage, make for good hearing.

In auditorium-type spaces, upholstered seats should be used whenever possible, to provide about the same absorption in the room, whether fully or partially occupied. Rear walls normally should be absorptive to prevent echo. Usually this provides enough absorption for acceptable reverberation.

In speech and drama spaces, or spaces for widely varying types of music, somewhat less reverberation may be provided by using movable draperies (retractable into pockets or chambers) to adjust the absorption within the room. In music spaces, the audience usually prefers considerable low-frequency reverberation.

In almost any space, reflections following quickly after and from approximately the same direction as the direct sound tend to enrich and reinforce the direct sound.

Divisible Spaces

Divisible spaces present difficult problems. Rarely are the spaces of the correct proportions or configuration in one of the conditions—divided or combined. Two tandem classrooms usually open into a poor large space, with ceilings much too low, and distance from front to rear too great.

Operable walls in auditoriums must have absorption on the side facing the stage, or they will produce serious echoes when the walls are extended.

Movable and operable wall performance is subject to many variables, in addition to the construction of the wall. (See Section III, Table 3-7, page 114.) Usually it is imperative to specify the *in place* performance of a wall (which may require a field test) to assure adequate isolation. An isolation rating of at least STC-30 is necessary for even minimal separation between simultaneously used spaces; STC-35 is better for ordinary classroom and conference rooms; and values in excess of STC-40 are normally imperative for audio-visual and auditorium dividers.

Except in small or noncritical spaces, get the recommendations of an acoustical consultant when designing divisible spaces.

Open Plans

Open plans present many difficulties (see particularly Section III, page 93, "Layout and Orientation of Spaces"). Some important design elements include:

1. Use much absorption, widely distributed (completely carpeted floors, much ceiling absorption with at least NRC .75 rating, and some wall absorption in most cases) to minimize reverberation and reflections.

2. Separate activity groups and noise sources as much as possible, and direct sound sources (teachers, audio-visual speakers, telephone operators, etc.) away from competing groups or persons.
3. Do not even attempt to locate very noisy equipment or spaces with critical privacy requirements in open plan areas. *Always* enclose them in separate rooms.
4. Use furniture, screens (preferably highly absorptive on both sides, with an STC rating of at least 20, and not less than 60 in [1.5 m] high), and other barriers to cut off sight and sound between activities.
5. Use masking to establish a steady, neutral background level; but *never* attempt to "drown out" high noise levels with masking. A well-shaped masking spectrum, not higher that dBA 47 in level, is usually acceptable and quite effective in adding to privacy; at higher levels it is intolerable to most people.
6. Try to provide *at least* 20 dBA separation between separate work stations, using a combination of the above elements. It is very difficult to provide more than 25 dBA separation in open plan spaces.

An open plan space is *not* simply a conventional building with the walls omitted. It is a highly special area, with many specialized requirements. While it is normally chosen to provide functional flexibility (which it can accomplish), it is also frequently thought to be "much cheaper" than enclosed spaces (a costly illusion which has embarrassed many designers and owners).

In general, it is folly to attempt to design open plan spaces without the assistance of competent, experienced acoustical consultants.

"Coupled Spaces"

Avoid deep, low-ceilinged spaces (under-balcony areas, deep transepts, etc.) coupled to the principal room volume. They "speak back" into the principal space, and their acoustics are often different from the principal space.

Poorly Proportioned Spaces

Whenever possible, avoid long, narrow spaces; rooms with large floor area and very low ceilings; and small rooms with very high ceilings. If the room "doesn't look right," it probably won't sound good, either.

Organ Chambers

Organ chambers and "mixing" chambers for instruments are highly critical spaces. If they must be used, *always* get professional assistance. Many fine instruments have been ruined by installing them in poor spaces.

SOUND REINFORCEMENT

Functions

Electrical sound reinforcement systems are used to:

1. Increase the "signal level" of the source.
2. Direct sound to the listener at a level high enough to be intelligible above the background.
3. Provide the output for audio-visual inputs of various types.

(**Note:** These functions are not identical with those of the so-called "intercom" and "P.A." systems.)

A good sound reinforcement system will provide amplified sound at adequate level, with fidelity and realism.

A system may provide "full-range" reinforcement (30 to 15,000 cps [Hz]) for music and similar sources; or it may be used to reinforce speech frequencies only (300 to 5000 cps [Hz]).

Reinforcement is usually necessary in the following spaces:

1. Legitimate theaters with more than 1000 seats.
2. Lecture halls with more than 300 seats.
3. Almost all gymnasiums, arenas, and large assembly halls.

Reinforcement systems are usually desirable even in smaller spaces; they are imperative as the output for audio-visual inputs.

Performance Criteria

A good reinforcement system can:

1. Provide realistic sound at levels up to 100 dB at distances up to 100 ft (30 m) from the speakers.
2. Direct sound with accuracy into the designated areas with uniformity of ±3 dB from the average SPL.

However, not much more than 9 dB of reinforcement (gain over the level of the natural source) can be provided without being detected. Ideally, the amplification ratio should be 1/1; that is, the level at the listener's ear should be about the same as that about 36 in. (1 m) (or normal distance from source to listener) away from the natural source. Therefore, the background level must be at least 10 dB (and preferably 20 dB or more) below the source level if the "ideal" amplification ratio is to be maintained. In very noisy spaces, the ratio must be increased, as required, to permit the amplified sound to "outshout" the background (by at least 10 dB, and preferably more). However, then the listeners will be aware of the system.

Speakers

Speakers should be located as near the source as possible, and arranged so that their signal *follows* the direct signal by a few milliseconds (*never* precedes the direct signal unless the direct signal is too faint to be heard at the listener's position). The central speaker cluster (or "high-level") arrangement permits this in most cases. Where distributed speaker arrangements ("low-level") must be used, electrical delays may be required in the circuitry to delay the amplified signal enough to permit it to follow the direct signal.

As illustrated in Figure 3-19, the most remote location covered by a speaker normally should not be more than three times as distant as the nearest location.

Speakers should direct their sound only into the listeners' areas, without "spraying" acoustic energy over reflective areas or into unoccupied space (to excite room reverberation or develop unwanted echoes). Speakers should not face one another, nor should their dispersion patterns cross or overlap appreciably. They should not direct their energy into operating microphones.

Use enough speakers in distributed speaker systems to cover all listening areas uniformly. A good "rule-of-thumb" for distributed speaker spacing is:

$$D = 2(H - 4)$$

where:

D = distance in feet between centers of speakers;
H = ceiling height in feet;

or:

$$D = 2(H - 1)$$

where:

D = distance in meters between centers of speakers;
$\dot{H}$ = ceiling height in meters.

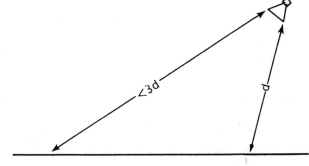

Figure 3-19 "Throw-ratio" for speakers.

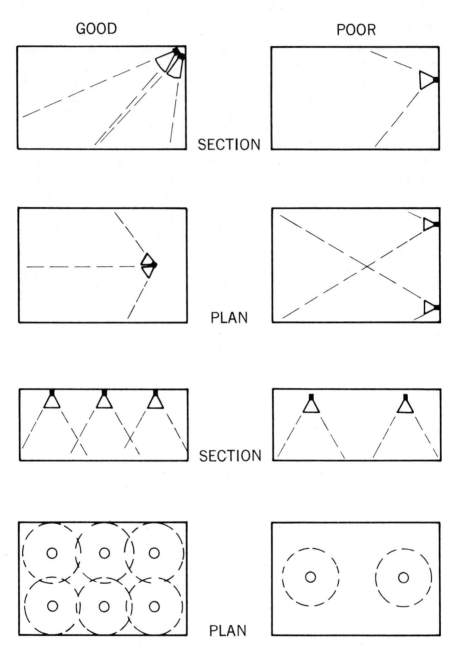

Figure 3-20 Speaker location and coverage.

Microphones

Use the best microphones which funds will permit. Top-quality condenser microphones provide the best fidelity and widest frequency range. Excellent dynamic microphones, with smooth response and wide range are usually lower cost and are acceptable for most purposes.

Microphones may be "omnidirectional," reinforcing sound from all directions; "cardioid," accepting essentially only the sound from the hemisphere which they face; or highly directional, accepting energy from narrow cones. Pattern should be chosen according to need for accepting or rejecting sound from various directions.

Source distance to microphone should be as short as possible (to provide the most favorable "signal-to-noise" ratio) and as constant as possible (to prevent "fading" as the source recedes or "blasting" as the source approaches the microphone).

Amplifiers

Use power amplifiers, preamplifiers, and line amplifiers, as required, to provide adequate signal level. Be sure that power ratings are real, long-term values ("peak power" ratings are usually at least twice actual steady output capacity of the unit). Amplifiers should have minimum distortion throughout the entire usable frequency range (30 to 15,000 cps [Hz], at least). Provide adequate channels for at least all anticipated inputs.

Controls

Whenever possible, provide means to adjust level of *each* speaker and to turn *each* speaker on or off, as required.

Provide level and on–off controls for *each* input, as well as for the overall system.

Impedance Matching Devices

Match impedances carefully, using the best quality transformers and similar devices. Cheap, high-loss transformers are always a mistake.

Accessories

Provide adequate microphone outlets for any anticipated requirement.

Use best quality, shielded microphone cable; and *nonshielded* speaker cable. Never run microphone and speaker cables in the same conduit or very close together.

Use absolutely *no splices of any kind* in any cable. Connect only at readily accessible terminal strips or at the equipment itself. Solder and wrap any pigtail connections at speakers.

Equalization

Use amplifiers and preamplifiers which permit "shaping" the frequency response spectrum by attenuating or boosting the low and high frequencies. This will permit correcting for room peculiarities.

For highly reverberant, acoustically difficult spaces, highly sophisticated, "custom-tailored" equalization is provided by using special filters to correct for strong room modes or other undesirable room responses. This is a highly specialized operation, which should be performed only by a skilled professional.

Remember: The room and the reinforcement system are elements of the same acoustical "instrument." They must be carefully fitted and tuned to one another. A sound reinforcement system is *not* hardware or an accessory.

A sound system will *not* correct bad room acoustics; rather, it may amplify and exaggerate them. The microphone is a monaural listening device which accepts indiscriminately whatever it "hears"; if noise goes in, noise comes out.

(Refer to Section II, page 68, "Sound Reinforcement," for a detailed discussion of sound reinforcement systems.).

Mechanical Equipment Noise and Vibration Control

Sound and vibration are associated with nearly all equipment containing moving parts, whether they are of the rotating, reciprocating, vibrating, or impact type. It may be:

1. *Inherent* in the operation, and essentially unavoidable; or
2. *Incidental* to the operation, and avoidable by good design or substantially eliminated by good practice (that is, it results from malfunction, improper adjustment, improper maintenance, etc.).

Table 3-10 (below) lists some of the more common equipment and the nature of its sound and vibration output. (See also Table 3-6, Section III, page 94.) Equipment typical of that found in heating, ventilating, and plumbing systems; typical production equipment; and examples of the more common noise sources in buildings are shown; and the nature of their sound and vibration output is described. In the table:

Type of motion = principal motion of major sound-producing element. (Usually, motions involve complex combinations of various types, with one type more important in generating the dominant sound.)

Type of sound = usual subjective description of sound produced. (Usually, the sound is a combination of several types with one type or frequency range dominant.)

"f" = fundamental frequency (in cps or Hz) of sound produced. (Highest peak pressure may occur at some higher harmonic of "f".)

TABLE 3-10   Mechanical Equipment Noise

| Source and Type of Motion | Characteristics of Sound or Vibration | |
	Inherent	*Incidental*
Rotating:		
Fans Blowers Turbines Centrifugal pumps Centrifugal compressors Centrifugal chillers	A tone of frequency: $f = \dfrac{\text{rpm}}{60} \times$ number of blades on wheel or impeller (and higher harmonics)	Aerodynamic "roar" (broadband) Dynamic imbalance with vibration frequency: $f = \dfrac{\text{rpm}}{60}$ and higher harmonics
—	—	—
Cooling towers	Fan noise and water splash	Same
—	—	—
Motors Generators	"Whine" of frequency: $f = \dfrac{\text{rpm}}{60}$ or some multiple	Same and cooling fan noise
—	—	—
Gears	"Whine" of frequency: $f = \dfrac{\text{rpm}}{60} \times$ number of teeth; or some multiple	Vibration with frequency similar to inherent noise. High-speed impact and sliding noise; broadband "grinding" and "screeching"
—	—	—
Bearings	"Squeal" of frequency: $f = \dfrac{\text{rpm}}{60} \times$ some multiple	Same

TABLE 3-10 (continued)

	Characteristics of Sound or Vibration	
Source and Type of Motion	Inherent	Incidental
Grinders	"Grinding" noise. Essentially broadband, often with a relatively strong pure tone of frequency: $f = \dfrac{rpm}{60} \times$ some multiple	Same
Saws Planers Routers Shapers	"Scream" of frequency: $f = \dfrac{rpm}{60} \times$ number of teeth, blades, or cutters	Same and "ring" of saw
Reciprocating:		
Internal combustion engines	A "roar" of frequency of firing rate: $f = \dfrac{rpm}{60} \times$ multiple of number of cylinders Exhaust noise	Cooling fan and pump noise; valve "clatter"; air noise Dynamic imbalance with vibration frequency: $f = \dfrac{rpm}{60}$ and higher harmonics
Compressors Pumps	Same Intake and exhaust noise	Same Pressure pulses in gas and fluid lines; frequency related to inherent sound frequency
Vibrating:		
Transformers Ballasts Rectifiers Light filaments	Relatively pure "tones" of frequency: $f = 2 \times$ cps of AC current or some harmonic	Sympathetic vibration of housings, casings, and attachments, some multiple of inherent sound
High-speed vibrators Vibrating conveyors	"Vibration" or "buzz" or "rattling"	Same

TABLE 3-10 (continued)

Source and Type of Motion	Characteristics of Sound or Vibration	
	Inherent	*Incidental*
Vibrators High-speed shakers	Essentially broadband	
—	—	—
Bells Buzzers Vibrating horns	Relatively pure tone; frequency related to method of generation	Same
Impact:		
Presses Hammers Shears Riveters Punches Tumblers Shake-out devices Low-speed vibrators Office machines Printing and duplicating equipment Accounting machines Print-out equipment	"Hammering, rattling, pounding, thumping, banging," etc. Essentially broadband	Vibration, broadband in frequency
Flow Noise:		
Airflow in ducts Fluid flow in pipes	"Rushing" or "flow" sound; relatively broadband	Same Cavitating pumps or supply/return imbalance create squeals and pulsations.
—	—	—
Valves and metering devices Throttling devices Dampers Flash tanks Orifices Nozzles	"Rushing, whooshing, swishing" type of sound Often a strong, almost pure "tone" or "scream" or "screech" Often very high frequency	Same Sound often travels long distances down ducts or pipes via walls or via fluid or gas. Strong pulsation or hammering when valves or throttling devices close or open

NOISE AND VIBRATION CONTROL

Choose Quiet Equipment

It is far simpler to avoid making noise than to eliminate it after it has been produced. Some equipment is inherently quieter than others. It is invariably more economical to choose a quieter, more expensive machine than to use a cheaper type which requires considerable additional noise and vibration control. For example, in an air-conditioning installation, the choice of chillers, in order of preference is:

First: Absorption Machines
Second: Centrifugal Chillers
Last: Reciprocating Chillers

It is often prohibitively expensive to use the cheaper reciprocating equipment because of the extra cost involved in providing adequate noise and vibration control for it.

Fan noise output, for a given capacity, varies markedly. Usually the well-made centrifugal type, particularly those with backward-curved, airfoil blades, are the least noisy. Large, high-speed propellor types are usually most noisy, since the tip speed becomes very high, even if the rpm is relatively low.

Low-pressure, high-capacity blowers are normally very noisy; they should be used with discretion. Gear pumps and other oil hydraulic pumps are often exceedingly noisy, even though they may be low-horsepower units.

Manufacturers of inherently noisy equipment often make available "quiet models" or noise and vibration control "kits" as "standard options." This is often a tacit warning that the equipment may be troublesome without such provisions. It is usually wise to buy the "silenced package" whenever possible.

Typical of such equipment are:

1. Hydraulic elevator pumps, which should be enclosed in absorptive housings, with built-in vibration isolation and pulsation dampers in the hydraulic lines.
2. Reciprocating refrigeration compressors, which normally require mufflers in the hot gas lines.
3. Lobed blowers, which nearly always need special mufflers, particularly on the suction lines.
4. Large, dry-type transformers, which are available with special housings, vibration isolation, and "quiet" models.
5. Large horsepower electric motors which may be purchased with special built-in cooling fans, at least 10 dB quieter than the "universal" fan.

Choose Operating Parameters Which Minimize Noise

The effect of operating parameters on machinery noise output is often enormous. For example, it may be tempting to increase the rpm of a machine or to

TABLE 3-10b   Effect, in dB, of Operating Parameters on Machine Noise[a]

Internal combustion engines	$10 \log_{10}$ horsepower ratio
	$30 \log_{10}$ speed (rpm) ratio
Fans	$10 \log_{10}$ horsepower ratio
	$10 \log_{10}$ pressure head ratio
	$50 \log_{10}$ rotational speed (rpm) ratio
Pumps	$17 \log_{10}$ horsepower ratio
	$40 \log_{10}$ speed (rpm) ratio
Gas flow	$80 \log_{10}$ velocity ratio (Mach 1 and higher velocities)
	$60 \log_{10}$ velocity ratio (velocities less than Mach 1)
	$30 \log_{10}$ pressure ratio (at velocities less than Mach 1)
Liquid flow	$60 \log_{10}$ velocity ratio (without cavitation)
	$120 \log_{10}$ velocity ratio (with cavitation)

[a]To determine the effect of changing an operating parameter (such as horsepower, rpm, etc.) on the noise output of a particular piece of equipment:
1. Determine the ratio of its present operating parameter to that of the proposed new parameter; for example, *double* the speed or *halve* the horsepower.
2. Insert this number (2 or $\frac{1}{2}$, for example) into the equation.
3. Add this value algebraically to the existing noise level of the machine to determine the new or changed level resulting from the change in operating parameter.

increase the flow velocity of water through pipes, but the penalty in increased noise may be prohibitive.

Table 3-10b shows the approximate effect, in dB, of modifying operating parameters. As is apparent, it may be much wiser to use a slower engine or pump and accept a small economic penalty in first cost than to speed them up and accept both energy and noise penalties.

Most equipment operates most efficiently when it operates most quietly. Noise tends to be an indicator of wasteful operating conditions.

Adjustment and Maintenance Procedures

Equipment should be operated at optimum performance speeds and capacities. Well-balanced, well-maintained fans, operating near peak efficiency, are usually much quieter than those operating under other conditions. Chillers with improper refrigerant charge or pressure or those operating under very light loading often are very noisy compared with their normal mode. Excessive noise and vibration usually indicate inefficient and damaging operation.

Equipment Location

Locate equipment in rooms or enclosures with thick, heavy walls, floors, and ceilings (or in other high transmission loss enclosures), as remote from critical areas as possible. The roof, top floor, or intermediate floors of buildings are

DON'T

Attach equipment rigidly to light, large panels or other surfaces.

DO

Mount equipment on massive, rigid parts of the structure.

DO

Separate equipment bases or equipment room floor from the structure.

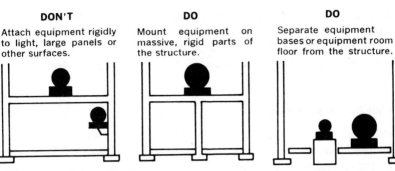

Figure 3-21   Minimizing structural transmission of noise and vibration.

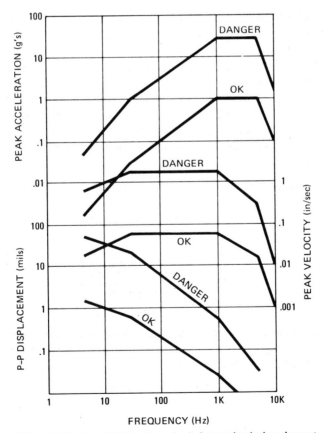

Figure 3-21b   Typical vibration criteria for mechanical equipment.

troublesome locations for heavy, noisy equipment. Either the basement, sub-basement, or a completely separate building may be preferable. To minimize problems, expose as few people and as little area as possible to noise and vibration. (See Table 3-5, Section III, pages 98–102, for mechanical room construction criteria.)

Rotating Equipment Imbalance

Well-balanced rotating equipment may exhibit less than one-tenth the vibration of equipment approaching dangerous vibration levels. (See Figure 3-21b.) This means that radiated noise from machine surfaces or attached panels may increase by as much as 10 to 20 dB as equipment condition deteriorates. Obviously, maintaining equipment is an effective noise control procedure.

Noise Radiation from Vibrating Panels

Panels, particularly those of sheet metal, glass, and similar hard elastic materials, readily set into vibration by vibrating sources attached to them, radiate noise efficiently into their surroundings.

Since it is relatively simple to measure the vibration level of a panel (with an accelerometer) and the airborne noise level in the near field of the panel (with a sound level meter), correlation between levels is feasible. This makes it possible to determine whether a vibrating element may be a significant contributor to excessive noise levels, as well as to unwanted vibration.

Some useful formulas, sufficiently accurate for an initial investigation of such problems, follow.

1. Airborne noise level in the near field of vibrating panel, radiating noise into the surroundings:

$$SPL = a - 20 \log_{10} f + 150 \text{ dB}$$

where:

a = acceleration in dB re 1 g;

$$\text{or } 20 \log_{10} \frac{\text{measured acceleration}}{\text{acceleration of gravity}};$$

f = frequency in cps (Hz).

2. Airborne noise level in the near field of a vibrating panel, radiating noise into the surroundings, measured with the A-weighted network for both the accelerometer and microphone amplifiers:

$$SL = 20 \log_{10} V + 146 \text{ dBA}$$

where:

V = measured velocity in m/sec;

SL = measured Sound Level, A-weighted.

3. Relation between acceleration, frequency and displacement:

$$G = \frac{4\pi^2 f^2}{386.1} D$$

where:

D = displacement in inches;
G = linear acceleration in g's.

4. Relation between frequency, velocity, and displacement:

$$D = \frac{V}{2\pi f}$$

where:

D = displacement in inches;
V = linear velocity in in/sec;
f = frequency in cps (Hz).

Acoustical "Treatment" of Mechanical Equipment

Even the best equipment, properly located, adjusted, and maintained may produce noise and vibration in excess of acceptable levels in many instances. Then the usual available options for "treatment" are:

1. Absorption of some of the sound.
2. Partial or total enclosure of the equipment.
3. Isolation of the equipment on resilient mounts or on independent bases or foundations.

Absorption alone is of very limited value. Applied liberally to the machine room, it will reduce mid- and high-frequency noise a few decibels. It will reduce room reverberation and it will tend to confine the noise to the vicinity of the machine; but it will do little for the operator exposed to the noise. If the machines are distributed throughout the space, it will do little to lower levels, although comfort and communication may be improved somewhat.

Partial barriers will provide some shielding for those exposed (except the operator; it may increase noise at his location). Total enclosure (when possible) will provide considerable reduction; even partial enclosure will usually help considerably. Absorption within the enclosure will further reduce noise outside of the enclosure.

Resilient mounting of the equipment is almost imperative to avoid "driving" the floor or the building structure, spreading the sound for large distances via the structure.

Damping vibrating panels or large surfaces attached to or adjacent to the machine will often reduce radiated sound. (Refer to Section II, page 78, for a detailed discussion of these various approaches to noise reduction.)

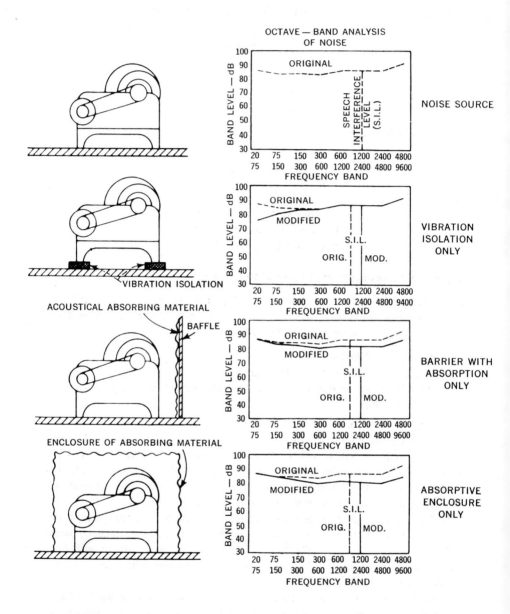

Figure 3-22a Effect of sound control measures. The improvement provided by any one of these measures alone is small—rarely adequate to eliminate a serious noise problem. (Courtesy General Radio Company.)

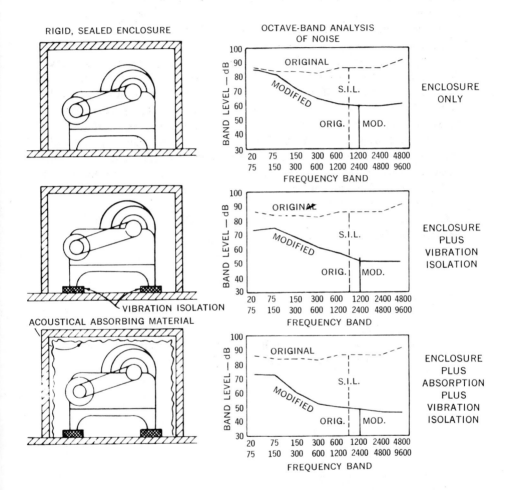

RIGID, SEALED ENCLOSURE

OCTAVE-BAND ANALYSIS
OF NOISE

ENCLOSURE
ONLY

VIBRATION ISOLATION

ACOUSTICAL ABSORBING MATERIAL

ENCLOSURE
PLUS
VIBRATION
ISOLATION

ENCLOSURE
PLUS
ABSORPTION
PLUS
VIBRATION
ISOLATION

Figure 3-22b Combinations of all types of sound control measures result in major improvements in all frequency ranges. Costs, severity of the problem, and amount of improvement required determine which approach or procedure is best. (Courtesy General Radio Company.)

Vibration Isolation

In general, all vibrating equipment should be resiliently mounted. Vibrating equipment includes rotating, reciprocating, and oscillating machines or devices of any type. In some cases, such as equipment on earth-supported foundations, sufficient isolation may be provided by eliminating structural connections between the foundation and the structure (refer to Figure 3-21). Usually, however, resilient mounting systems are necessary.

Usually it is advisable to design for at least 90% isolation (10% transmissibility); for critical spaces (apartments, some laboratories, etc.), 95% or more isolation is advisable; for industrial plants and for basement installations (on a floor on the earth) of small equipment, 85% isolation may be acceptable. (Refer to Section II, page 58, for a discussion of transmissibility in resiliently mounted systems.)

In most instances, use the *lowest* frequency of the slowest moving part as the "driving" (or "forcing") frequency to be isolated; this will usually assure adequate isolation for the higher frequencies which may be present. For example, a system with a 1725 rpm motor, driving a 600 rpm fan, should be isolated to protect against the 600 rpm frequency.

The "natural" (or "resonant") frequency of a resiliently mounted system may be calculated from the formula:

$$f = 3.13 \sqrt{\frac{1}{d}}$$

where:

f = frequency in cps (Hz);
d = static deflection of the spring in inches under the load imposed (determined by the stiffness of the spring);

or:

$$f = 5 \sqrt{\frac{1}{d}}$$

where:

f = frequency in cps (Hz);
d = static deflection of the spring in centimeters under the load imposed (determined by the stiffness of the spring).

The natural frequency of the system should be kept to less than $\frac{1}{3}$ of the driving frequency to provide appreciable isolation (and less than $\frac{1}{5}$ for critical installations).

The required static deflection for resilient mounts may be determined directly from the graph, Figure 3-23, following.

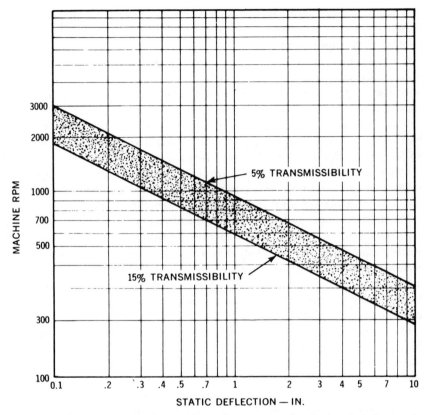

Figure 3-23 Required static deflection for resilient mounts. Required static deflections for resilient mounts can be determined on the graph by using the slowest-moving component in the mounted system as the "machine rpm." For "critical" installations, use 5% Transmissibility curve (or above); for ordinary installations (basement machine room, low-horsepower equipment) 15% Transmissibility curve is usually acceptable.

In practice, rarely is the calculated isolation achieved, since it assumes a simple single degree of freedom (up and down motion only), a perfectly elastic spring, and an infinitely stiff base. However, with proper precautions and mounting system design, the formula is sufficiently accurate for most work.

The dead-load deflection of the base supporting the resiliently mounted system (whether it be structural members or the earth) reduces the effective isolation of the resilient mounts and increases the transmissibility of vibration.

If the natural frequency of the mounting system coincides with that of the supporting base, resonance may occur. This will not only vitiate the isolation, but it may cause dangerous amplification of the vibration in the supporting structure.

A good "rule of thumb" is to provide at least three times (and preferably four or more times) as much deflection in the resilient mounts as the dead-load deflection of the supporting structure.

For the usual types of equipment, mounted on reasonably heavy structural slabs or slabs on grade, the deflection of the supporting structure is small and need not be taken into account. However, large, heavy, slow-moving equipment may produce enough unbalanced force to require examining carefully the deflection and frequency of the supporting structure or the earth beneath it. For most reasonably dry, well-compacted earth, the natural frequency lies well above 20 to 30 cps (Hz); for some wet, "marshy" types of soil, it may be as low as 8 to 12 cps (Hz). Long-span, highly flexible trusses or beams may have large dead-load deflections and low natural frequencies of vibration. It is quite possible to create a serious resonance condition if the isolation system is not carefully designed.

The deflection of each support in a resiliently supported system must be essentially identical, or the system may develop a complex "rocking" or "twisting" motion rather than a simple, predictable vertical motion. This is one of the most seriously neglected points in resilient mount selection. It is very common to see an unsymmetrical machine resting on four identical mounts, with one nearly unloaded, and one or more "bottomed" and not functioning at all; or with the machine rocking violently.

Insofar as possible, the mounting system should be designed to keep the center of gravity of the mounted mass within the plane of the supporting surfaces of the resilient mounts. This is particularly important for reciprocating equipment, equipment likely to produce large unbalanced forces, and systems supported on very flexible mounts (more than $\frac{1}{2}$ in. [1.2 cm] static deflection).

The motor (or engine) and the driven machine *must* be on a common, rigid base. It is imperative that the driving and driven parts of a system *never* be separately mounted on resilient supports or serious damage may result to one or the other part.

Obviously, design or choice of resilient mounting systems is not a casual procedure nor one that should be undertaken by the naïve. Experienced acoustical consultants and competent manufacturers' representatives are usually capable of providing dependable and safe recommendations.

Inertia Blocks To reduce motion, lower the center of gravity, minimize the effects of unequal weight distribution of the supported machine, and stabilize the entire resiliently mounted system, a heavy, rigid "inertia block" is often used as the base for the equipment. Then the entire assembly is supported by resilient mounts under or attached to the inertia block.

Generally an inertia block should be at least 6 in. (15 cm) thick, and very stiff and rigid to avoid significant flexure in any direction. Usually it consists of reinforced concrete poured within a structural steel channel frame to which are attached structural members to carry the load to the resilient mounts. When many pad-type mounts are distributed under the inertia block, the channel frame may not be necessary.

The inertia block should usually be equal in weight (or greater than) the weight of the supported equipment. (See Table 3-11, for required weight for various types of equipment.) When the mass of the supported equipment is enormous (for example, very large centrifugal compressors), there may be no need for additional mass in the form of inertia blocks; a rigid frame to support the entire assembly may be sufficient. This is particularly true if the moving parts are small compared with the total mass of the unit.

Housekeeping Pads   In addition to adding mass to the floor beneath any piece of equipment, a housekeeping pad serves to keep the resilient mounts and mounted equipment up off of the floor, so that the mounts will not be fouled or the system "short-circuited" by debris. Whenever feasible, provide a minimum clearance of 1 to 2 in. (2.5 to 5 cm) between the top of a housekeeping pad and the bottom of the supported inertia block or equipment frame, to permit easy visual inspection of the free space between them. For small equipment on pad mounts, this is less important, since inspection usually is simple.

Resilient Mounts   Mounts must be chosen to provide the required static deflection; sufficient stability to avoid excessive movement (vertical or lateral); and sufficient damping to minimize excursions of the mounted system during start-up or changes in operating conditions.

Because of the damping inherent in all materials, they never act as completely elastic materials; hence their isolation efficiencies never reach the calculated amount of an ideally resilient material. Further, springs and resilient pads are subject to "wave effects" which permit much more energy transmission at some harmonics of the forcing frequency than the theoretical transmissibility calculations predict.

Spring mounts should be used for static deflections of $\frac{1}{2}$ in. (1.2 cm) or more. A ribbed-rubber or similar pad, at least $\frac{1}{4}$ in. (6 mm) thick, should be provided under the base of the spring mounts to avoid transmission of high-frequency vibrations which might travel down the spring. The ratio of horizontal to vertical spring constants should be about one, whenever feasible. Large-diameter, unhoused, free-standing springs are desirable; other mounts should incorporate provisions for snubbing only extreme horizontal or vertical displacements, such as occur during start-up of a large fan or pump. Such snubbing must not be counted on to provide full-time stability for an inadequately designed spring.

For static deflections up to 0.4 in. (1 cm), rubber-in-shear mounts may be used. However, they should not be used when natural frequencies less than 6 cps (Hz) are required in the mounting system.

For static deflections of $\frac{1}{4}$ in. (6 mm) or less, ribbed-rubber, rubber-and-cork "sandwich" pads, and similar elastomeric pads may be used. However, it is not safe to depend upon them to provide natural frequencies under 10 cps (Hz); and it is highly likely that, with age, they will stiffen sufficiently to increase the natural frequency to somewhat higher values. Pads must be of proper size, area, and thickness to carry the superimposed loads within their rated capacities and static deflections.

(**Note:** Properly designed glass-fiber pads, of proper density, fiber type, and bonding material act essentially as elastomer pads. Their limitations and performance characteristics are similar to elastomeric materials.)

For very low frequencies (natural frequencies under 2 cps [Hz]), air mounts are usually required. They are excellent for this purpose, but their expense and maintenance problems usually limit their use to special applications.

It is imperative that *any* mount be loaded properly so that its static deflection does, in fact, equal the required deflection. Springs too lightly loaded will not provide either the deflection or isolation required. Pads too lightly loaded will not isolate adequately; when too heavily loaded, they may crush or fatigue quickly.

There are many excellent sources of mounts of all types. Always obtain the manufacturer's recommendations and recommended loading and installation procedures for any mount.

Mounting System Requirements Resiliently mounted equipment is not a collection of parts, but an integral system. Requirements vary with type, weight, and horsepower of equipment. The following table provides typical recommendations for the usual mechanical equipment found in most buildings (other than manufacturing plants). (See Table 3-11.)

Shock Isolation

Shock isolation is an extremely complex procedure. The impact of large forging hammers, punch presses, shears, and similar equipment often involves enormous forces and large amounts of energy. Usually the operation requires that the anvil or base which holds the parts being worked on must be rigid and massive. This is fortunate, since the mass provides the inertia to stabilize the entire system. The motion of the hammer (or moving mass of the system) must be stopped quickly, within a short distance, dissipating its energy as work done on the part. However, considerable energy is imparted to the machine foundation.

The usual problem is to slow the rate at which the excess energy is applied to the foundation and to dissipate as much as possible within the mounting system.

The isolation techniques are similar to those for vibration isolation, but the differences between the two procedures are important. Equipment motion

TABLE 3-11 Mounting System Requirements

| Equipment | Inertia Block | | Resilient Mounts | | |
	Required[a]	Weight	Type	Minimum Static Deflection[b] (in.)	(cm)
Reciprocating:					
Pumps					
Compressors	Yes	2W[c]	Springs	1	2.54
Engines					
Centrifugal:					
Pumps					
Compressors					
Blowers			Pads		
Less than 3 hp	No	—	Rubber-in-shear or springs	$\frac{1}{4}$	0.64
Greater than 3 hp	Yes	2W[c]	Same	$\frac{1}{4}$	0.64
Fans:					
Less than 10 hp	No	1W[d]	Same	$\frac{1}{4}$	0.64
Greater than 10 hp	Yes	1W[d]	Springs	1	2.54
Cooling Towers:					
Centrifugal	No	—	Springs	1	2.54
Propellor	No	—	Springs	$2\frac{1}{2}$	6.35

[a] Requirement may be relaxed slightly for small fans and low-horsepower centrifugal equipment supported on a floor on grade or a slab or block on the earth.
[b] Choose deflection from Figure 3-23. Must be at least three, and preferably four or more, times the dead-load deflection of the supporting structure.
[c] Multiple of weight of supported equipment, including any directly supported piping, etc., and any vertical hydraulic thrust which might be applied to the mounted system. Mounts, also, must provide for this total weight or force.
[d] Multiple of weight of fan and motor plus scroll and other directly connected sheet-metal work, up to the flexible connection where the scroll joins any ductwork.

Note: Special instruments and similar sensitive apparatus often require similar mounting systems. Depending upon required stability and isolation requirements, the above systems may be used; or more stringent requirements may demand air mounts or special procedures.

usually must be very limited, and the resilient mounts must be much stiffer than would normally be the case for vibration isolation mounts. For example, the same pads which are used for vibration isolation may be loaded only about one-half as heavily under shock loading. Often layers of hardwood timbers are inserted between the resilient pads and the masonry foundation for the equipment; the timbers crush slightly and protect the masonry from damage resulting from the huge dynamic forces involved.

Knowledge of soil and subsoil conditions, allowable loading, proximity of structures subject to damage, etc., are essential. The design of shock isolation systems is usually best left to the experts.

Heating, Piping, Air Conditioning, and Electrical Systems

(**Note:** For a detailed discussion of noise and vibration control in heating, ventilating, and air-conditioning systems, refer to the "Sound Control" chapter of the current *ASHRAE Guide*. This book will not attempt to duplicate the information contained in the *Guide,* nor will it cover in detail the procedures outlined there. Rather, it will discuss the general principles involved and some "rules of thumb" and good practices in designing systems.)

GENERAL

The mechanical and electrical equipment associated with these systems in most modern buildings is a source of noise and vibration. Generally, equipment should be housed in rooms built of heavy, tight construction, as remote from occupied or critical spaces as possible. (See Table 3-5, Section III, page 98, for mechanical room construction criteria.) The sound control principles and procedures described in preceding sections apply to this type of equipment also.

FAN AND DUCT SYSTEMS

Fans should be mounted as described previously (Section III, page 169, "Mounting System Requirements"). It is usually most economical to choose the quietest fan capable of supplying the required airflow. Fans produce high noise levels, even when their horsepower is small.

Operate fans "on the curve" to assure maximum efficiency and minimum noise. Use the slowest fan velocity consistent with other system requirements. Propellor-type fans should rarely be operated at fan tip speeds in excess of 6000 or 7000 fpm (1850 to 2150 mpm) where noise control is important. (See Table 3-10b, Section III, page 158, for the effect of operating parameters on noise output.)

In general, it is best to eliminate as much noise as possible as near to the fan as possible; the major exception to this is in high-velocity systems, where additional provisions are always necessary.

Fan Housings and Plenums

Thick lining (2 to 4 in [5 to 10 cm]) in fan housings and plenums is a very effective means of attenuating fan noise. Lined housings also reduce the noise level within the machine room. Very large housings for high-capacity fans often must be mounted on resilient supports to minimize transmission of "rumble" to the floor. This is particularly true in high-velocity, high-pressure systems when fan horsepower approaches 100 hp. (See Figure 3-24, for one type of mounting system.)

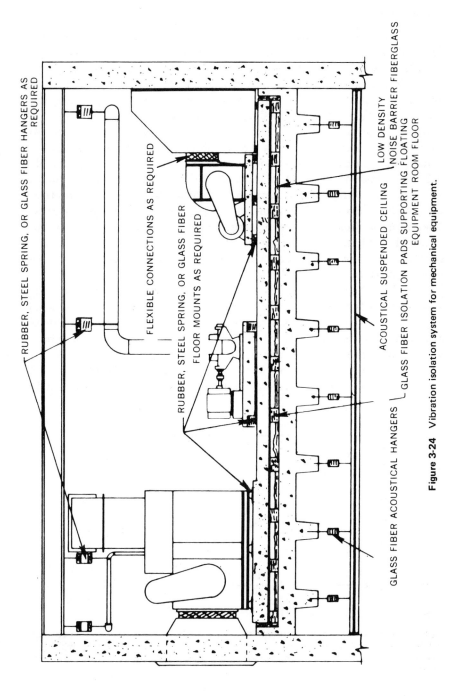

Figure 3-24 Vibration isolation system for mechanical equipment.

Duct Systems

Fan noise is transmitted readily down duct systems—supply *and return* ducts alike. The duct system is, in effect, a huge, continuous "speaking tube" unless properly treated.

Flexible connections should be provided between resiliently mounted fans and any ducts leaving them. Ducts should be adequately braced and of sufficiently heavy gauge metal to prevent excessive "drumming" or vibration. Standard ASHRAE recommendations are usually adequate for low-pressure, low-velocity systems. High-velocity systems may require substantially heavier construction.

The flow of air ducts is also a source of sound. Even at low velocities (under 2000 fpm [600 mpm]) the flow is usually no longer laminar, and turbulence occurs at turns, irregularities, sharp cross-sectional changes, and throttling devices. At high velocities (3000 fpm and over [900 mpm]), flow noise becomes a major problem. Good aerodynamic design of the entire duct system is advisable even for low velocities, but it is imperative for high-velocity systems. (Good duct system design is discussed in the *ASHRAE Guide* and in numerous books on ventilating system design.)

Figure 3-25 shows the effect of throttling devices in a duct system. Obviously, a minimum of throttling or similar restrictions should be used. Dampers should, in general, be as far as possible from the outlet (grille or diffuser). It is difficult and expensive to eliminate the 10 to 15 dB of acoustic energy which may be introduced into the system at a damper.

In general, use the lowest flow velocity feasible; use as little throttling as possible; and keep flow velocity (average linear face velocity) through grilles and dif-

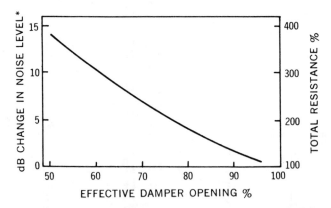

Figure 3-25 Effect of damper closure (*PWL, dBA, or NC). **(Note:** Dampers are occasionally closed down deliberately to generate "masking" noise. However, this is a complex procedure, and it often creates system balancing problems. It should be done only as a last resort.)

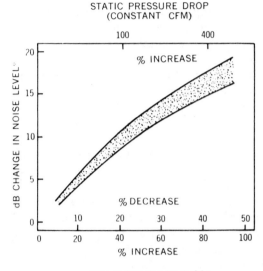

Figure 3-26 Effect of changes in operating conditions on performance of typical diffusers (*PWL, dBA, of NC). Static Pressure Drop increases approximately as the square of the increase in CFM through the diffuser. With a given CFM, and an increase in Static Pressure Drop, the noise rating tends to increase as if CFM had increased. Thus, for a doubling of pressure drop, noise increases as if CFM had increased about 40%. This does not appear to hold for very large changes in pressure or for unusual damper arrangements. Diffuser manufacturers usually supply "Sound Ratings" for their units; the curve shows the effect of changes in operating conditions and how they affect sound output.)

fusers to a minimum (under 350 fpm [100 mpm] in critical spaces). Figure 3-26 (above) shows the effect of change in airflow through diffusers.

Manufacturers' noise data for diffusers are usually based upon optimum ducting to the diffuser. However, grilles or diffusers inserted directly into the side of a duct, on a very short stub, or very near a damper or throttling device, are generally much more noisy than those located at the end of a reasonably long branch duct.

Variable volume systems, with continuously changing flow through the entire system, produce constantly changing noise levels. It is important to size grilles and diffusers according to the acceptable noise level at *maximum* flow.

Again, it is very difficult, particulary in the room into which the air discharges, to eliminate 10 to 15 dB; it is better not to generate the noise.

Duct lining, properly used, will attenuate fan noise in duct systems quite effectively. Where required, it should be at least 1 in (2.54 cm) thick and of at least $1\frac{1}{2}$ lb/ft^3 (24 kg/m^3) density (or heavier), coated, and of long-fiber, incombustible material.

When duct runs—*both supply and return*—are less than 50 ft (15 m) long to the nearest outlet or return grille, it is highly likely that duct lining alone is not adequate to reduce fan noise to acceptable levels (unless the duct runs are unusually tortuous). In such cases, prefabricated duct mufflers may be required. Duct mufflers (or silencers) are a special form of "splitter" for air discharge or intake of fans, where flow velocity does not exceed approximately 6500 fpm (2000 (mpm). For higher flow velocities, special constructions are normally required. Special, acoustically absorbent louvers can be used in intake or discharge openings where noise transmission to the environment must be minimized; and aerodynamically shaped, absorbent turning vanes may be used in large duct elbows. Such devices can provide from 5 to 30 dB attenuation (depending upon the frequency of the sound).

Mufflers are not particularly effective at low frequencies. A lined plenum, at least 6 ft (2 m) long, with lining at least 2 in (5 cm) thick, immediately downstream of the muffler is often adequate to supplement the high attenuation provided by the muffler in the middle and high frequencies; such a plenum between two short mufflers is particularly effective.

Confusion and misunderstanding about procedures for calculating and predicting ventilating system noise persist among architects and engineers. Many professionals apparently assume that all that is required is to choose (from manufacturers' catalogs) diffusers which, under design operating conditions, will produce noise levels meeting the required NC-criteria for the spaces they serve. This assumption might be valid *if* diffuser noise were the sole (or major) source of sound for the particular duct run. Then the sound level in the space could be controlled by manipulating the volume flow and the pressure drop through the diffuser. Rarely, however, does this simple condition exist.

In low-velocity systems, with long, lined ducts or with adequate muffling (in both supply and return runs), it is likely that diffuser noise will control the noise level of the ventilating system in the spaces served. However, this holds only if fan noise has been adequately attenuated, dampers or throttling devices are not excessively restricted, and the actual flow noise of the air in the duct produces levels well under the diffuser noise levels. The increasing use of high-velocity, high-pressure systems; unducted or plenum return or supply; ventilating light fixtures; and other unusual arrangements make these assumptions naïve and, on occasion, disastrous.

In large or complex systems, or where critical acoustical conditions are encountered, a complete system noise analysis (as outlined in the *ASHRAE Guide*) is advisable. A quick "rule-of-thumb" check for simple systems involves calculating the attenuation of the system at 125 and 250 cps (Hz) only, since fan noise is usually highest at these frequencies, and they are the most difficult to attenuate. For example:

	125 cps (Hz)	250 cps (Hz)
Fan Sound Power Level (obtain from manufacturer)	90 dB	87 dB
Attenuation of bare (unlined) housing	−4 dB	−4 dB
	86 dB	83 dB
50 ft (15 m) *lined* duct (@ 0.25 dB/ft @ 125 cps [Hz]; 0.33 dB/ft @ 250 cps [Hz])	−13 dB	−16 dB
	73 dB	67 dB
Two *lined* elbows (@ 2 dB each @ 125 cps [Hz]; 3 dB each @ 250 cps [Hz])	−4 dB	−6 dB
	69 dB	61 dB
"Room Effect" (Not less than 5 ft [1.5 m] from outlet in reasonably "soft room)	−9 dB	−9 dB
Sound Pressure Level in occupied space	60 dB	52 dB

This is equivalent to approximately NC-45, acceptable for an open plan office or similar space.

Had the duct been *unlined*, the Sound Pressure Level in the room would have been about 10 to 12 dB higher, requiring a duct muffler or silencer approximately 7 ft (2 m) long in the duct run to attenuate the noise as much as the lined duct. An acoustically lined fan housing would provide at least 5 to 8 dB more attenuation than the unlined housing; and a tortuous duct run or a run with several branches before the outlet in question would further reduce the Sound Pressure Level at the outlet.

The usual hardware and accessories (turning vanes, extractors, mufflers, fittings, etc.) may actually become serious noise producers in high-velocity systems; furthermore, regenerated noise is so universally a problem that standard lining or muffler arrangements are of limited value in such systems. Special duct with special lining and facing material (such as double-wall round or oval duct, with an interior perforated metal liner covering 1 in. [2.54 cm] or more of dense duct liner) is usually required; and heavily lined mixing boxes (at least $\frac{3}{4}$ in. [2 cm] lining), followed by a large, long, lined plenum (as much as 10 to 15 ft [3 to 5 m] long) or a similar length of flexible duct, is usually necessary to attenuate regenerated airflow noise at each outlet. When airflow velocities exceed

2000 lineal feet per minute (600 mpm), the problem may arise; at 3000 fpm (900 mpm) or above, it almost surely will arise.

It is highly advisable to use the services of a qualified acoustical consultant when undertaking the design of any of the special systems discussed above.

Almost as troublesome as the fan and duct system noise is the problem of flanking and "cross talk" via ducts and the openings and interconnected volumes they use or create. An unlined duct is a large, effective "speaking tube." A continuous duct, serving several rooms, may permit sound to travel readily from room to room, vitiating all of the architectural sound isolation provisions.

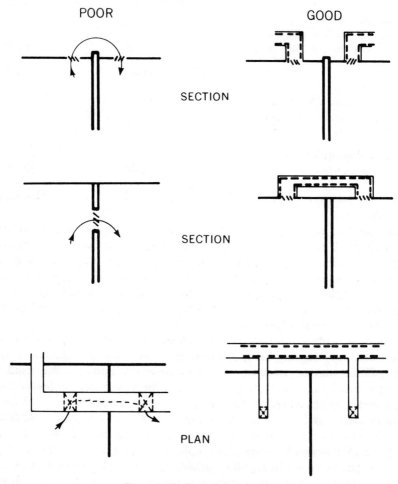

Figure 3-27 Typical flanking paths.

Typical flanking paths include:

Door undercuts	Continuous plenums above ceilings
Relief grilles	Duct shafts
Transfer grilles	
Ventilating light fixtures	Continuous supply plenums serving strip diffusers

Figure 3-27 shows some typical flanking paths and some measures to avoid or correct problems.

If a sound barrier must be penetrated, the duct must not provide an easy path for sound transmission, or it will become a major "leak." Particularly serious are the relief and transfer grilles and door undercuts in high transmission loss constructions. They must not be used where good sound isolation is required. A sufficiently long length of lined transfer duct, with at least two right-angle turns, will normally prevent an acoustical leak; in very critical spaces (radio studios, etc.) special ducting is usually imperative.

A continuous plenum above the ceilings of adjacent spaces may effectively couple all of the spaces unless extreme care is taken to avoid flanking. This is particularly true when ventilating light fixtures, distributed ceiling return grilles, and similar devices are used.

Occasionally, continuous plenums are created inadvertently by the continuous cabinet or enclosure for under-window convectors or similar devices. Often the partition between spaces merely abuts these enclosures rather than penetrating them to close off the opening between spaces. This is certain to cause trouble, since it provides an enormous leak between rooms.

Another common and serious leak often occurs at the perimeters of ducts (and pipes, conduits, etc.) where they penetrate walls and floors. It is important that these joints be small and carefully cut or formed; and that they be packed with mineral or glass wool (or similar fibrous material) and the visible joint on both sides of the barrier caulked with flexible caulking material.

Miscellaneous Ventilating Equipment

All of the preceding principles apply to the small, individual fans, fan-coil units, wall and roof-mounted ventilators, exhaust fans, unit heaters, and similar equipment. They are sources of sound and vibration; even though small in size and capacity, they should not be overlooked or ignored.

Resilient mounting provisions are built into good small-fan equipment. Sound-absorptive curbs are available for roof-mounted ventilators. Although still rather noisy, window air conditioners and fan-coil units are becoming quieter; and actual sound test data are available for some units.

It is important to remember that such small equipment is often located either directly in occupied spaces or very near to them. Therefore, it is not safe to ignore it because it is "small."

PIPING SYSTEMS

Pumps, compressors, chillers, and similar machines are always connected to piping systems. Therefore, in addition to choosing quiet equipment and mounting and housing it as discussed in the preceding pages, it is necessary to consider the attached piping as a transmission path. Sound (and vibration) may travel via the walls of the pipes or via the fluids (or gases) contained in them. This latter path is often ignored, although it may be the more important of the two.

Usually, most noise and vibration results from the pressure pulses introduced into the system by the impeller or the pistons of the pump or compressor, but fluid flow noise may also cause problems.

Piping rigidly attached to resiliently mounted equipment may "short-circuit" the mounting system and transmit vibration to the structure; therefore, some type of resilient or flexible attachment is required, in *both* the discharge and suction lines. For small pipes serving small equipment, it is occasionally sufficient to insert three successive elbows into the lines as near the machine as possible. Then the pipe lines should be suspended from resilient hangers. Such an arrangement permits freedom of movement in all three axes (as in Figure 3-28).

Flexible Hangers

Prefab hangers with springs, rubber-in-shear, or glass-fiber pads inserted to eliminate any solid connection between pipe and structure to which the hanger is attached are highly advisable for pipe lines of 2 in (5 cm) diameter or over (and

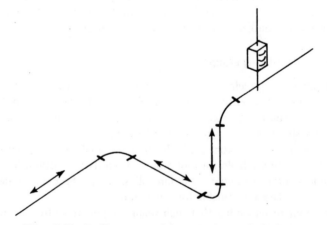

Figure 3-28   Resilient connection system for small pipelines.

for high-velocity ducts and very large air ducts suspended near critical spaces). Such hangers are advisable for approximately 100 pipe diameters of run from the machine.

However, such arrangements do nothing to eliminate or minimize transmission of pressure pulses from the machine; nor do they minimize flow noise.

Flow Noise

Flow noise can be minimized by using low flow velocities—under six lineal feet per second (2 m/sec) whenever possible; by using smooth, long-radius, well-made fittings; and by avoiding anything which creates turbulence. (See Table 3-10b, Section III, page 158, for the effect of flow velocity on noise generation.)

Pipe "lagging" (dense, thick insulation) or a thick layer of resilient insulation (Fiberglas or urethane foam) covered with a heavy, resilient sheet (leaded vinyl or several layers of roofing felt, etc.) effectively reduce radiation of flow noise from pipes.

Resilient Connectors

Resilient pipe connectors minimize vibration transmission, prevent "short-circuiting" the machine mounting system, and absorb and smooth out the pressure pulses which might otherwise be transmitted down the line via the fluid. It has been found that the ability of the connectors to "balloon" slightly makes them much more effective in minimizing transmission of pulsations. Therefore, only the elastomer type (not metallic or braided types) are recommended. The standard spool-type expansion joint (with a built-in arch or arches) is very effec-

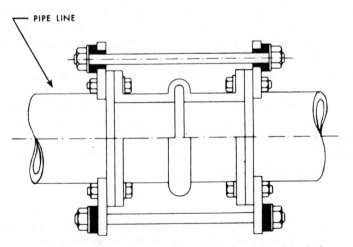

PIPE LINE

Figure 3-29 Flexible pipe connector with isolated restraining bolts.

tive as a flexible connector if the restraining bolts are isolated with elastomer bushings to avoid metal-to-metal contact.

For optimum results, the flexible connector should be in a *horizontal* segment of pipe, as near the pump as possible. For large lines, it is better to use two connectors in the discharge line—one as near the pump as possible, followed by a long-radius elbow, and a second connector following the elbow. The elbow should be supported firmly, either on the pump inertia block, on a resiliently mounted stanchion, or from a resilient hanger system which restrains its motion without "short-circuiting" the resilient systems. The pipe line, immediately beyond the resilient connector, must be similarly restrained.

Mufflers

Internal combustion engines, reciprocating compressors, and some types of blowers create considerable intake and exhaust noise. An adequate commercial muffler (usually a three-stage type) is usually required to minimize such noise.

Valves

Valves tend to create turbulence, particulary when throttled down. The resulting "squeal" or "scream" often travels for great distances down the pipe, as well as creating high noise levels near the valves. Steam-pressure-reducing valve stations are particularly prone to this, especially when large pressure reductions are effected in a single stage. A by-pass arrangement to handle low flow through more open valves is advisable. In critical installations, a pipe spool of a least 10 pipe diameters in length should be installed in the piping immediately downstream of each steam-pressure-reducing valve. If the valve proves to be objectionably noisy, a commercial muffler can be installed in place of the pipe spool. Commercial "diffractors" (a type of perforated tube or cone which straightens out the turbulent flow and makes it more laminar) is often effective in reducing such noise, too (5 to 15 dBA attenuation is possible).

The inertia of a mass of fluid flowing in a pipe may be substantial. If flow is stopped by a quick-closing valve, such as a metering valve, a loud "water hammer" sound may result. In large lines, this may actually be damaging. Slow-acting valves are highly advisable in all lines; when they are not acceptable, it is imperative that some form of pulsation control (air chamber, water hammer arrester, accumulator, etc.) be inserted in the line to receive and store the energy and to return it slowly to the line.

Drain Pipes

Drain pipes leaving pumps or similar equipment should be physically separated from drains or pipes which are attached to any part of the building structure. Any detail acceptable under local plumbing codes is usually acceptable acoustically.

Flash Tanks

Flash tanks should be installed in a mechanical equipment room remote from critical spaces, or they should be placed within a sound-isolating enclosure.

Air and Steam Discharge

Air, steam, and gas discharge, through orifices or nozzles, into occupied spaces may be very objectionable and actually damaging to the hearing of those exposed. Low-cost commercial mufflers are available to reduce such noise significantly (10 to 30 dBA); their use is always advisable.

ELECTRICAL EQUIPMENT

Electrical Connections to Resiliently Mounted Equipment

A long, flexible loop of flexible conduit, at least 20 diameters in length, should be used between rigid electrical conduit and any resiliently mounted equipment.

Transformers

Transformers (with their rigid connecting bus bars, if any), some very large electromagnets, and similar electrical equipment should be supported on resilient pads or mounts capable of providing approximately 0.20 to 0.25 in (5 to 6 mm) static deflection when loaded.

Motors

Large electric motors often produce high "windage" noise; and the noise of the cooling fan in air-cooled motors can reach high levels. "Mutes" for motors are available to attenuate such noise without interfering with ventilation or cooling; they can provide as much as 10 dBA attenuation.

Elevators and Escalators

Most elevator noise results from vibration produced by the motors, pulleys, cables, hydraulic pumps, and similar equipment. Methods of reducing such noise have been discussed previously.

A commonly overlooked problem is that of the impact and vibration of the moving elevator car against the guide rails or of escalator trusses rumbling as the stairs move. It is advisable to attach guide rails to massive, stiff parts of the structure only (such as the edges of floor slabs—*never* to the walls) and to use resilient pads between the bearing plates and the structure. Likewise, resilient pads under the bearing plates of escalator trusses will significantly reduce vibration and rumbling.

Ballasts for Fluorescent Lights

Whenever possible, ballasts should carry a sound rating of "A" or "B." In spaces where, for example, sound recordings are to be made, it is questionable whether fluorescent lights should be used at all. If used, the ballasts should be remotely mounted.

Miscellaneous Equipment

Some rectifiers and some light filaments "hum" or "buzz" objectionably. Rectifiers may be located remotely from critical areas; but noisy lights cannot be used in critical spaces (such as some TV studios, etc.).

OTHER MECHANICAL EQUIPMENT

Many pieces of plumbing, heating, and mechanical equipment are serious noise sources. Typical offenders are:

> Boilers
> Burners
> Kitchen equipment
> Water coolers

The noise and vibration sources within them consist of one or more of the pieces of equipment discussed previously; and sound control measures for this equipment are identical with those for such equipment.

Cooling Towers and Air-cooled Condensers

As previously discussed, cooling towers and condensers should be resiliently mounted to minimize vibration transmission. Since such equipment is normally located out-of-doors, either on the roof of a building or on the ground, adjacent to the building, it can become an environmental noise problem.

The noise of the intake and discharge air of cooling towers and the fans of air-cooled condensers can reach very high levels. Shielding, enclosing, or muffling such equipment is becoming increasingly necessary, particularly in urban areas.

If at all possible, such equipment should be designed to operate at two levels—maximum (with all fans operating at full speed) and reduced (with fans operating at reduced speed or with a portion of the fans turned off) as cooling requirements change. Normally this permits quieter operation at night, as the heat load is reduced and neighborhood noise becomes less acceptable.

Other noise control options include mufflers for air discharge and intake; special housings for any compressors contained in the condensers; damping for radiating panels; and even "fences" or other acoustical barriers to provide partial enclosure or shielding.

OTHER SOUND CONTROL DEVICES

A host of "standard" devices are commercially available for various mechanical noise and vibration control purposes. Typical are flexible shaft connectors, resilient mounting rings for bearings and pillow blocks, etc. Most of them contribute to quiet performance and should not be overlooked.

Sound Fields and Sound in Enclosures

GENERAL

Up to this point, we have dealt principally with known and relatively "typical" problems of sound, noise, and vibration control in buildings usually designed by architects and their engineering consultants. The recommendations include standard good practice and time-tested solutions, as well as established "rule-of-thumb" suggestions for avoiding known problems.

Increasingly, however, architects and engineers must deal with problems which, if not unique, may not be susceptible to "cookbook" solutions. Chief among such problems are those associated with industrial plants and with environmental noise control. Such problems usually require a measure of true engineering analysis.

In Section II, pages 79-82, "Sound Fields and Sound in Enclosures," a discussion of the mathematics of sound propagation in various types of spaces is given in some detail. The principal formulas are repeated here for easy reference.

In Section III, pages 160-161, some formulas for correlating radiated noise with vibrations generating it are discussed. With this information, it should be possible to handle some of the ordinary industrial and environmental noise problems. However, many such problems require the services of competent, experienced consultants, and should not be undertaken by the novice.

SOUND FIELDS

Sound radiating freely from a *point source* radiates spherically into the surrounding space; hence, in its near field (out-of-doors or before any reflective surfaces affect it), its intensity drops by 6 dB as the distance from the source doubles.

Radiation from a *line source* reduces in intensity by 3 dB as the distance from the source doubles.

Most noise sources, though, tend to be random, with nonuniform radiation from many surfaces—large, flat planes; small, variously shaped components; highly directional cavities, etc. As a rule, it is necessary to make measurements in many locations around a machine to determine its *sound power* output and how the acoustic energy radiates from it. From these same measurements, the *directivity* of the source can be deduced; it is *not* safe to assume uniform, spherical, hemispherical, or line radiation from a large machine! (See Section III, Table 3-3, "Effect of Distance" for data on attenuation with distance for machines located out-of-doors.)

The principal formulas for relating Sound Pressure Level to source Sound Power Level in nonreverberant space are:

$$\text{Sound Power Level} = SPL + 20 \log_{10} r + 0.7 \text{ dB}$$

where:

SPL = the Sound Pressure Level in dB;

r = distance from point source to measurement position in feet;

or:

$$\text{Sound Power Level} = SPL + 20 \log_{10} r + 11 \text{ dB}$$

where:

SPL = the Sound Pressure Level in dB;

r = distance from point source to measurement position in meters.

$$SPL = PWL + 10 \log_{10} Q - 20 \log_{10} r - 0.7 \text{ dB}$$

where:

SPL = the Sound Pressure Level in dB;

PWL = the Sound Power Level in dB;

r = distance from point source to measurement position in feet;

Q = directivity;

or:

$$SPL = PWL + 10 \log_{10} Q - 20 \log_{10} r - 11 \text{ dB}$$

where:

SPL = the Sound Pressure Level in dB;

PWL = the Sound Power Level in dB;

r = distance from point source to measurement position in meters;

Q = directivity.

(**Note:** SPL is often written as L_p; PWL is often written as L_w.)

SOUND IN ENCLOSURES

As might be expected, sound contained within an enclosure is not free to propagate indefinitely according to the above formulas. The energy which reflects from enclosing surfaces is added to the sound field and affects it according to these formulas:

$$SPL = PWL + 10 \log_{10} \left[\frac{Q}{4\pi r^2} + \frac{4}{R} \right] + 10.5 \text{ dB}$$

where:

r = dimensions in feet;

$R = \dfrac{\alpha S}{1 - \alpha};$

S = total area of the room surfaces;

α = average absorption coefficient of the surfaces at a given frequency;

or:

$$SPL = PWL + 10 \log_{10} \left[\frac{Q}{4\pi r^2} + \frac{4}{R} \right] + 0.2 \text{ dB}$$

where:

r = dimensions in meters;

$$R = \frac{\alpha S}{1 - \alpha};$$

S = total area of the room surfaces;

α = average absorption coefficient of the surfaces at a given frequency.

All of the terms following "PWL" in the above two formulas can be lumped into a quantity sometimes called the "room effect" (actually, the distance plus room effects). For large rooms and reasonable distances from the sound source, this "room effect" is a negative quantity, and the Sound Pressure Level (SPL) is lower than the PWL by several dB. For typical industrial spaces, the "room effect" can be taken from the following table:

TABLE 3-11b   Room Effect in dB

Distance from Source		Hard, Reverberant	Soft, Absorptive
feet	meters	Spaces	Spaces
10	3	– 12 dB	– 17 dB
20	6	– 15	– 22
30	9	– 17	– 25
40	12	– 18	– 27
50	15	– 19	– 29
100	30	– 20	– 34

Note: Never use this table for calculation of small enclosures (such as machine enclosures), machine housings, fan housings, or similar spaces. It is intended only as a quick, reasonably accurate means of forecasting the SPL in a work space if machine or source Sound Power is known.

Never use this table for calculations where the directivity (Q) of the source is high. A horn or any peculiarly directional source will "focus" the sound sharply.

In the near field of a very large sound source, there may be little attenuation within the first 10 ft (3 m) from the surface of a source, even in a totally absorbent field.

If a source is completely enclosed within an enclosure, the radiated sound energy must either escape or "build up" to where the escaping energy just equals the generated energy. Within a hard, reflective enclosure, the Sound Pressure Level at a machine may be much *higher* than it would be if there were no enclosure. An operator within the enclosure would be exposed to higher levels than if there were no enclosure. Further, however good the sound attenuation (STC or transmission loss) of the enclosure, it will accomplish little shielding if there is little or no acoustical absorption within it.

Applying acoustical absorbents to the interior surfaces of the enclosure will cause some of the energy within it to be absorbed, and the noise level within the enclosure will build up only by the amount not absorbed. For example, if the average absorptivity of all surfaces within the enclosure were only 10%, the level inside could build up by 10 dB; if average absorptivity were 50%, the build-up would be only 3 dB.

(For the effect of typical small enclosures, see Section III, Figures 3-22a and 3-22b, pages 162–163.)

ENVIRONMENTAL NOISE CONTROL

While man's "environment" includes all of his surroundings, modern usage has corrupted the word to imply the out-of-doors (just as "ecology" has become one of the most abused words in our vocabulary). Reluctantly, we must adapt to this misuse of the word, lest we be accused of "speaking in tongues." In the past, architects and engineers (particularly engineers) often neglected the effects of their work on the acoustical environment outside of their buildings, factories, planes, and automobiles. No longer is this possible, since legislation, as well as popular opinion, has made professionals responsible for the noise their works produce.

The problem is essentially that of the "sound fields" surrounding the buildings and machines—the propagation of sound and vibration from the sources into the neighborhood.

Noise Sources

The principal noise sources affecting the environment today are:

Transportation and traffic—planes, trains, trucks, buses, automobiles and motorcycles

Construction activities—pile drivers, jackhammers, pavement breakers, concrete mixers, hammering and pounding, sawing, blasting, crushing, and allied activities, as well as the transportation vehicles involved

HVAC equipment—fan intakes and discharges, compressors, motors, pumps, transformers, etc.

Sporting and entertainment activities—human shouting and cheering, music and loudspeakers, shooting, race cars, motorcycles, motorboats, and the like

Industry—factories, mining, and similar production facilities

They are all associated with essential and necessary activities (at least most people would agree with this assessment), so that they cannot simply be eliminated.

Noise Control at the Source

Since legislation is mandating maximum noise levels for most transportation and construction equipment, and legal limits for the noise output of a host of other equipment are being studied, the specifier or purchaser can do little, other than to:

> Choose the quietest type available;
> Operate it in its quietest mode;
> Keep it well maintained;
> Restrict hours of operation.

For example, cooling towers, located outside where they might become a nuisance at night, can utilize two-speed motors, reducing fan speed during the cooler evening hours and lowering the noise output by as much as 10 dBA; or air-cooled condensers can operate only a portion of their fans at night when the heat load is reduced.

Aircraft and airport noise problems are so specialized and complex that only the most expert (and tough and patient) professionals should attempt to handle them. Separate branches of governments dispute endlessly over who has jurisdiction over what aspects of air transportation and where the public's best interests lie (or ought to lie). Federal regulations, governing the maximum noise levels of *new* aircraft may provide some future relief, but airline economics will determine when and if such equipment is ever purchased or operated. Currently, proximity to airports and flight paths may govern where and whether a housing project, school, or hospital may be built, particularly if it involves any Federal funds. Whenever possible, the designer must take into account the intrusive and almost uncontrollable nature of aircraft noise, and he must choose sites and constructions compatible with the uses of the building and its site.

Traffic noise will probably continue to be a problem, particularly in urban areas, but even in suburban and rural areas near major thruways. Efforts at highway planning and zoning can possibly minimize, but never "solve," this problem. Currently, governmental agencies attempt to forecast traffic flow and density and to establish various maximum noise levels at arbitrary distances from traffic lanes as design criteria.

Sporting and entertainment activities are "special" in that they involve the mores and customs of a people. Only when they clearly become nuisances can much be done to control their noise. Athletic contests can be conducted within large stadiums or enclosed field houses; and their sound reinforcement systems need special design care so that only the spectators (and not the entire neighborhood) hear the amplified sound. Shooting and similar activities are usually located far enough away from normal living areas (or are totally enclosed in special shooting ranges) not to interfere significantly with the peace and quiet of the public; little more can be done (other than to outlaw them completely).

Race tracks should always be located far from residential neighborhoods, particularly if night racing is planned. It is very difficult to contain the sound of powerful racing engines within any reasonable fence or barrier.

Noise Control Along the Propagation Path

While sound propagation out-of-doors in rural areas can be forecast reasonably well with the formulas previously described, unusual wind and weather conditions may have significant effects, with downwind levels as much as 5 to 10 dB higher than the formulas would predict. Unusual terrain and ground cover, on the other hand, may provide substantial excess attenuation over what the mathematics forecast. Propagation in urban areas, particularly where large buildings line the streets, is very difficult to calculate. For unusual sites and potentially critical exposures, special studies are always advisable before making major decisions concerning the acoustical effect of a noise source on its surroundings.

Noise control approaches for environmental noise problems are almost identical with those previously discussed. Sources must be properly located to minimize exposure, shielded, enclosed, or otherwise "treated" to prevent excess energy transmission to the surroundings.

Barriers—solid, heavy, and high—are used to interrupt direct transmission (slat fences, decorative screens, and trees are almost useless). Along highways, solid fences and earth berms can provide up to 10 dBA attenuation. It is usually wise to apply absorptive materials to the side of the screen facing the noise source to absorb as much energy as possible, lest the barrier merely transfer the problem to another area by reflection. Masonry screens and partial enclosures (often open top and one side), especially those made of slotted acoustical block, are highly successful, providing as much as 20 dBA attenuation.

When air movement through the barrier is imperative, acoustically absorbent louvers can provide up to 10 dBA attenuation.

Shock and vibration transmission can cause serious problems in the vicinity of forging hammers, rock crushers, large hammermills, and similar equipment. These problems are usually of such magnitude that only the most experienced practitioners should attempt to handle them.

Noise Control at the Receiver

Occasionally, environmental noise problems require noise control at the receiver, as well as at the source, since it may be impractical or even impossible to provide sufficient reduction at the source or along the propagation paths. Typical of such approaches are:

> Double-glazing the windows of surrounding buildings;
> Gasketing and sealing doors and openings;
> Providing additional sound insulation for roofs and walls;
> And even providing masking sound in the affected buildings.

In some instances, acquiring land and buildings for a considerable area adjacent to the noise source may be the most practical and economical means of solving the noise problem. (See Section III, page 194, "Legal and Medical Problems," for a discussion of criteria and regulations.)

INDUSTRIAL NOISE CONTROL

Because most industrial noise problems are those of "sound in enclosures," they are included in this section. Today there is such a wealth of published information on industrial noise control that no attempt will be made to provide an exhaustive treatment of the subject.

Most industrial noise problems and their solutions are similar to those associated with any building. The sound, noise, and vibration control approaches are basically the same, too. Usually though, they deal with sources which are much louder and more annoying—they may even be damaging to hearing or dangerous to health. Often they involve the operating facilities, medical, and safety departments of the clients, as well as their architects and engineers.

In contrast to the acoustical refinements of a performing space, sound control in a factory usually means reducing noise or vibration to a safe or tolerable level. Some of the more common industrial noise problems are discussed in the following pages of this section.

Noise Sources

(Refer to Section III, pages 153-156, "Mechanical Equipment Noise and Vibration Control.")

The most troublesome noise sources in industry include:

> Air, steam, and gas jets
> Cutting, grinding, and machining processes
> Impact processes
> Flames and burners
> Gears, bearings, and rollers
> Magnetostrictive devices
> Vibrating, oscillating, and reciprocating parts

Because they normally involve much more energy than typical equipment for ordinary buildings, their problems tend to be more serious and difficult to solve. Further, in most cases, it is not possible simply to shut them up in a room and leave them; often the equipment is out in the midst of many workers and is attended constantly by one or more operators.

A particularly annoying and unnecessary noise source in industrial plants is the noisy, almost unintelligible paging system, with cheap, improperly located horns (and usually far too few) blaring at levels well over 100 dBA, usually employed to reach one or a few key personnel who could be reached better with other means of communication.

Noise Control at the Source

All of the principles previously discussed for mechanical equipment noise and vibration control apply to industrial noise problems.

In addition to choosing quiet equipment, consider substituting a different process for the noisy process. For example, use forging presses rather than hammers; substitute scarfing for chipping; consider chemical milling rather than conventional milling; etc. Such an approach is limited, particularly when existing equipment represents a substantial investment, but it should not be overlooked.

One principle always valid in equipment noise control is to reduce power consumption as much as possible. Noise generation is nearly always directly proportional to the power applied to the operation. Any modification to tools, procedures, or equipment which reduces power consumption is almost certain to reduce noise.

Nozzles or open jets, widely used in industry, can be silenced significantly by reducing pressures to a practical minimum (30 lb/in.^{2} [200 kPa] for air discharge to atmosphere, for example), and by using discharge or exhaust mufflers specifically designed for this purpose. In addition, it is advisable to use automatic valves or similar devices to limit discharges to only the necessary part of the work cycle, rather than permitting continuous discharge. Noise reductions of 10 to 20 dBA are possible with such procedures.

Furnaces with properly adjusted burners, well-muffled air intakes, and proper breeching and stacks are always quietest; and electric furnaces are invariably much quieter than those with burners.

Vibrating conveyors, those with bare metal rollers, and the types which employ bottle-to-bottle or can-to-can impact to move parts are always noise problems. Newer, more efficient, and much quieter types are available for almost any use today. They can reduce noise levels by more than 20 dBA.

Bare metal work surfaces, particularly checker plate and similar sheet metal panels, "ring" under impacts. They can be quieted with the damping materials previously discussed; or elastomeric, plastic, or even wood sheets can be substituted for them or applied to them.

Noise Control Along the Propagation Path

After exhausting reasonable source noise control options, it may be advisable to work on the noise propagation paths.

Noisy equipment and processes should be isolated as much as possible, rather than scattered throughout the work space; this concentrates and minimizes propagation paths.

A major noise radiation "path" is often the stock being processed. Long boards being run through a planer, tubes being swaged, structural members being cut, etc., are efficient radiators of sound. Damping their vibrations with hold-downs, elastomeric rollers, or similar devices; or enclosing them completely (both input and output ends) may be advisable.

Completely enclosing the noisy machine or the noise-producing parts of the equipment could be effective, acoustically, but it is rarely feasible. However, considerable enclosure, with well-designed baffling or other means of reducing noise transmission through necessary openings for incoming stock and outgoing finished parts, is often practical. It is not within the scope of this book to cover the many variations of such enclosures, but the section on "Case Histories" (Section III, pages 213–226) describes some successful designs.

Screens and barriers, fixed or movable, of rigid or flexible material with an STC rating of not less than 20, can be used to shield operators from noise sources or partially to enclose noise sources. Such screens must be at least 72 in (1.8 m) high and long enough to extend at least 48 in (1.2 m) beyond each end of the noisy machine to provide significant shielding—from 5 to 15 dBA. Full-height wing walls, not less than 48 in. (1.2 m) long at either end of the screen can provide as much as 3 dB additional shielding, as can a horizontal or an angled panel at the top of the screen. Applying absorptive materials to the source (noisy) side of the screen is usually advisable, not only to reduce levels within the shielded area but to minimize reflections from the screen to other areas.

Vibration isolation and damping techniques previously discussed at length can minimize structure-borne transmission and radiation of noise. Radiation of flow noise from duct and pipe walls can be reduced by 5 to 10 dBA by "lagging" (applying dense thermal insulation) or wrapping them with viscoelastic or resilient layers (dense glass fiber or urethane foams) covered with heavy films (leaded vinyl, thin sheet lead, etc.)

Noise Control at the "Receiver"

Often the most economical and effective noise control method is to isolate the operator from the noise. This is particularly true if a single operator is exposed to a very large source (such as an automated machine line) or to many surrounding sources. It may be simple to provide either a small, totally enclosed booth or a partial booth (three sides and a top, for example). Normally the interior of such a booth should be highly absorptive; and viewing panels should provide at least as much attenuation as the panels themselves. From 5 to 25 dBA (or even more) protection is feasible with such structures, often at a fraction of the cost of other approaches.

Administrative Controls

A method of noise control, currently referred to as "administrative," consists of controlling the hours of exposure of operators so that cumulative, long-term exposure (8 hours a day for 40 hours a week, for example) is limited. This may mean rotating operators between noisy and quiet operations during the work period; scheduling noisy operations to a single or limited number of brief periods during the day; or it may involve limiting the operator's exposure to his own operation and controlling the cumulative time of his own noise generation to less than the permitted maximum.

Rarely does a "hands on" operation, attended by a full-time operator, produce noise for more than a minor fraction of the work cycle (some studies indicate that less than 20% of the cycle involves excessive noise in many such operations). Thus the operator, protected from adjacent noise sources, and exposed only intermittently to excessive levels of his own operation, may not experience excessive cumulative exposure.

In practice, this approach may be one of the more successful and economical.

(See "Case Histories," Section III, pages 213-226, and Table 3-14 "Cost/ Benefit Analysis of Industrial Noise Control Options." See Section III page 194, "Legal and Medical Problems," for a discussion of criteria and regulations.)

Personal Protection

For a discussion of personal protection as a noise control option, refer to "Legal and Medical Problems," Section III, pages 198-200.

Legal and Medical Problems

As discussed previously, the subjective effects of sound and vibration are complex and not completely predictable. This is particularly true when we attempt to evaluate noise within the framework of medicine and law. Common questions include:

1. When does noise become a nuisance?
2. What is a "safe" or "suitable" acoustical environment?
3. What is the effect of noise on productivity?
4. When does noise become a hazard to persons or property?

There are no simple nor complete answers to any of these questions. At best we can suggest some reasonably well-established criteria and danger signs. Beyond that, only additional research and experience (and judicial decisions and medical discoveries) can provide answers.

In previous sections certain background levels for various activities were suggested; and it was pointed out there is a significant level which most people consider "unpleasant" or "uncomfortable," and above which verbal communication is almost impossible. Significantly, it has been established that long, continuous exposure to levels above that level does tend to damage the hearing of most persons exposed. Therefore it seems reasonable to establish some sort of benchmark at that level.

HEARING CONSERVATION CRITERION

Overwhelming evidence suggests that long, continued exposure to noise levels in excess of 85 dBA is likely to degrade the hearing of a large percentage of humans. As a result, a well-established criterion has been widely used in sound control. Figure 3-30 shows the permissible Sound Pressure Level in octave frequency bands for continuous exposure to noise. This does *not* mean that exposure to sound levels in excess of the criteria will always cause hearing damage, but it is reasonably safe to assume that below these levels there is little likelihood of damage. Neither does this mean that such levels are comfortable or acceptable for all activities. In fact, such levels are much too high for comfort; and they make many activities very difficult, particularly those requiring easy verbal communication.

Above the criterion, as levels increase, exposure time before damage becomes increasingly brief. At about 120 dB, exposure should be limited to three to five minutes a day.

Because of the human, medical, and legal implications of the problem, it is highly advisable to seek expert advice whenever levels in excess of the criterion are encountered.

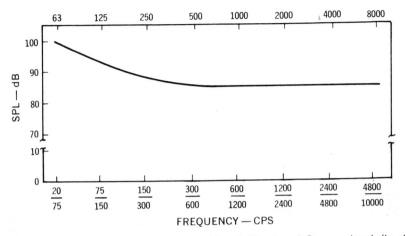

Figure 3-30 Hearing conservation criterion. Permissible Sound Pressure Level (in dB re 0.0002 microbar) in octave frequency bands for continuous exposure to noise.

Industrial Noise Exposure Criteria

A widely used criterion for long-term exposure (8 hours a day, for 40 hours a week) for industrial workers is 90 dBA, with a halving of the permissible exposure time for each increase of 5 dBA in level.

Since cumulative exposure to noise above certain levels is widely believed to account for the rate and severity of permanent hearing damage, some authorities believe that the "doubling rate" should be 3 dBA, since 3 dB represents a doubling of the energy level. Other authorities believe that 85 dBA should be the base level (rather than 90 dBA), above which the doubling rate applies.

A tentative standard, proposed for United States workers, is the OSHA table shown on page 196 (Table 3-11c).

VIBRATION CRITERIA

Beyond the discussion on page 15 of Section I, it is difficult to establish criteria for vibration. What people accept in a vehicle, they will not tolerate in their homes, offices, or shops. Certainly levels which damage structures, disorient or nauseate humans, or cause equipment and instruments to malfunction are unacceptable.

Rarely will complaints or problems arise when levels of vibration remain under −60 dB re 1 g (that is, one-thousandth the acceleration of gravity).

Damage to well-built structures is rarely found at levels below one-tenth the acceleration of gravity. (NOTE: This is not necessarily true when shock or impact are involved.)

TABLE 3-11c Industrial Noise Exposure Criteria

Steady Level Noise

Level (dBA)	Time (hr-min)	Level (dBA)	Time (hr-min)
85	16-0	101	1-44
86	13-56	102	1-31
87	12-8	103	1-19
88	10-34	104	1-9
89	9-11	105	1-0
90	8-0	106	0-52
91	6-56	107	0-46
92	6-4	108	0-40
93	5-17	109	0-34
94	4-36	110	0-30
95	4-0	111	0-26
96	3-29	112	0-23
97	3-2	113	0-20
98	2-50	114	0-17
99	2-15	115 (max)	0-15
100	2-0		

Noise of Varying Level

Daily dose (D) of unity must not be exceeded.

$$D = \frac{C_1}{T_1} + \frac{C_2}{T_2} \cdots + \frac{C_n}{T_n}$$

where

C_n = actual duration of exposure (hrs);
T_n = noise exposure limit (hrs) from table above.

Impact Noise

Must not exceed 140 dB Peak Sound Pressure Level.

dB SPL	Impacts/Day
140	100
130	1,000
120	10,000
110	100,000

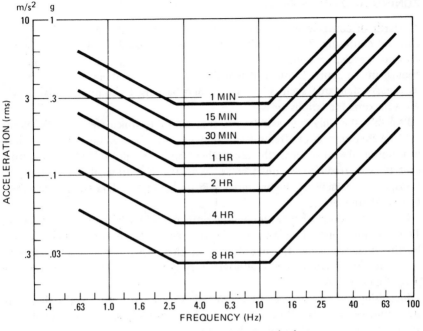

Figure 3-31 Vibration exposure criteria.

The problem is much too complex for simple "standard" or "rule-of-thumb" recommendations: the advice of experts is advisable in any important project or problem.

Industrial Vibration Exposure Criteria

Permissible exposure to vibration is far from a settled matter, but international standards organizations have proposed the levels and exposure times indicated in Figure 3-31 for seated or standing workers.

EFFECT OF NOISE ON PRODUCTIVITY

Little or no incontrovertible data exist on this subject. The research to date is so contradictory and confusing that it is rarely helpful. Almost certainly productivity is affected by the acoustical environment, but it is difficult to demonstrate this with objective evidence. The background levels suggested in previous sections are reasonably safe; they have evolved from long experience and many measurements in production plants and areas.

ZONING AND NOISE ORDINANCES

A dearth of court decisions and legal precedents make this entire subject difficult and confusing. Zoning performance standards, based upon objective Sound Pressure Levels, have not been on the books long enough or tested frequently enough in the courts to make safe any comments about them. Some cities have used naïve standards, such as "50 dB"; others have used a contour similar to NC-40 as maxima for various areas.

Probably the only safe assertion is that anything which clearly modifies the existing environment or affects it adversely may be adjudged unacceptable. Thus, in traditionally noisy areas, the acceptable standards will probably be less stringent than in traditionally quiet areas. Near manufacturing plants, airports, or transportation facilities, higher levels will probably be tolerated than in suburban, residential areas.

For short periods of time, and within normal working hours (usually 7 A.M. to 11 P.M.), noisy operations usually are permitted by the authorities and tolerated by the citizenry. Thus, the pile drivers, jackhammers, and air compressors required for heavy construction, although very noisy, are accepted as a necessary part of modern life.

Increasingly, however, people are refusing to accept noise which they consider unnecessary. The "sonic booms" of supersonic planes, for example, are apparently above the threshold of human acceptability.

The common law concept of "nuisance" probably remains the best protection for citizens against increasing and unnecessary noise. Noise, by definition, is a subjective thing—"unwanted" sound, whatever its character. Thus, the judgment of a jury of our "peers" continues to be our best safeguard against problems which still defy simple, objective formulation or definition.

Environmental Noise Criteria

Slowly a pattern of acceptable environmental noise limits is emerging. Rational or not, it is probably the type of consensus which evolves in situations as complex and subjective as "optimum" or "acceptable" criteria for human happiness and welfare.

Some criteria frequently found in the United States are shown in Table 3-11d following.

PERSONAL PROTECTION MEASURES

When it is necessary that humans be exposed to hazardous noise and vibration levels during their normal activities, some form of personal protection is advisable.

The airlines, for example, have instituted what is apparently a successful program of ear protection for persons exposed to jet engine noise. The earmuffs

TABLE 3-11d   Typical Environmental Noise Criteria

Location		Maximum
Residential, in vicinity of airports	NEF	30
	L_{dn}	65
	CNR	100
Residential, typical outside	dBA	55
typical inside	dBA	45
100 ft (30 m) from centerline of outer lane of express highway	dBA	70
Property line of industrial plant	dBA	60
Extreme boundary of major airport	PNdB	113

(**Note:** See Table 3-2b, Section III, page 90, for definitions of various criteria terms.)

carried by all ramp personnel are now as commonplace as safety goggles and hard-toe shoes in industry. Well-fitting earmuffs can provide as much as 30 dB protection in the critical frequencies. Even good earplugs are very helpful, although not as good as muffs. Ordinary absorbent cotton offers only a little protection, although wax impregnation appears to improve performance somewhat.

Experience indicates that protective devices for the ears are usually acceptable for short term, intermittent exposure, but considerable doubt exists about their effectiveness for continuous, long term protection. Ear muffs tight enough to be truly effective are uncomfortable; workers almost invariably relieve compression by springing the head band or by other expedients which significantly reduce the shielding afforded. Many workers object strongly to inserting anything into their ears, particularly well-fitting plugs, and they develop ingenious means to obtain comfort, including shortening or otherwise modifying the plugs until they are almost useless. Probably disposable, easily molded, impregnated fibrous plugs are as "practical" and comfortable as any device, and they probably provide as much protection as any other device actually *worn*.

It is advisable to plan on not more than 10 dBA protection from any of the available ear protectors worn continuously for long periods (8 hours a day, for example).

Well-planned programs of frequent and regular relief from exposure minimize the danger of hearing damage or physical injury.

When it is impractical to enclose the noisy equipment or otherwise isolate the noise and vibration from the equipment, it is often possible to remove or isolate the workers from direct exposure to damaging levels. For example, it is frequently possible to locate a well-designed enclosure or booth near the equipment; the workers may spend most of their time supervising or operating equipment from the safety of the booth. For those unavoidable excursions from the booth, workers may wear earmuffs.

In those industries where exposure to high noise levels is a normal part of the environment, a program of audiometric tests and ear examinations is advisable. By continually monitoring noise and hearing, possible damage can be minimized.

A well-designed, properly supervised, continuously monitored hearing conservation program, with frequent and regular audiometric tests (as well as ear protectors) is *not* a cheap substitute for reducing noise to safe and acceptable levels. In the long run, particularly with the likelihood of large personal injury and compensation claims, it may prove to be the most expensive solution to noise problems.

Tests and Measurements

Tests and measurements are means of making objective evaluations, establishing standards and criteria, and specifying performance requirements. Increasingly, architects, engineers, designers, inspectors, safety directors, and others with responsibility for man's environment must make at least preliminary measurements or perform some simple tests in order to make intelligent decisions.

The principal tests and measurement procedures used in applied acoustics are discussed briefly in the following paragraphs.

LABORATORY TESTS

The four principal laboratory tests used in sound, noise, and vibration control are:

1. Standard Method for Laboratory Measurement of Airborne Sound Transmission Loss of Building Partitions; ASTM Designation E 90-75 (and Standard Classification for Determination of Sound Transmission Class; ASTM Designation E 413-73).
2. Tentative Method of Laboratory Measurement of Impact Sound Transmission Through Floor–Ceiling Assemblies Using the Tapping Machine; ASTM Designation E 492-73 T.
3. Standard Test Method for Sound Absorption of Acoustical Materials in Reverberation Rooms; ASTM Designation C 423-66.
4. Geiger "Thick-Plate" Decay Test for damping rate of damping materials.

These tests are discussed in Section II and Section III. They are normally performed by independent testing laboratories with carefully standardized and controlled facilities.

FIELD TESTS

Usually, only two tests of building component performance are conducted in the field:

1. A "Simplified Field Sound Transmission Test" which involves measuring the Sound Pressure Level in the *source room* in dBA and in the *receiving room* in dBA, and referring to the difference in dBA between these two levels as a Field Noise Reduction. A new ASTM Test Method (E 597-77 T) has recently been published. It is based upon a method proposed in 1971 by the author, using the following procedure:
   a. Locate a "pink noise" (equal energy per frequency *band*, such as an octave-band or third-octave band) source in the room on one side of the test parti-

tion, approximately two-thirds the length of the room away from the test partition and facing *away* from the test partition. (See Figure 3-32.)

A noise generator capable of providing levels of 100 dBA or higher is advisable. In an emergency, one or more large, noisy industrial vacuum cleaners with their hoses removed are a reasonably acceptable "source."

b. With a sound level meter (Type I or II), measure the Sound Pressure Level, in dBA, in the source room at a point approximately one-third the distance down the long diagonal of the room from the test specimen. (If feasible, establish precisely 100 dBA at this point to simplify calculations.)

c. At the corresponding one-third point in the receiving room on the opposite side of the test specimen, measure the Sound Pressure Level, in dBA, *using the same sound level meter.*

Be sure the level in the receiving room with the sound source operating in the source room is at least 6 dBA (and preferably 10 dBA or more) higher

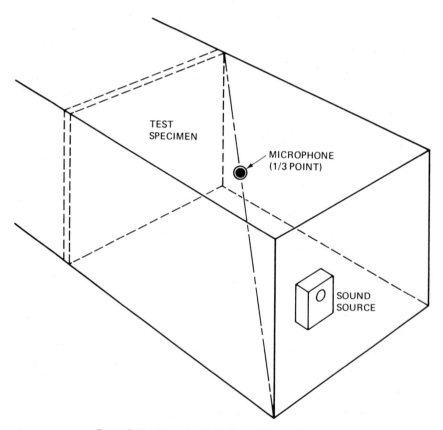

Figure 3-32 Simplified Field Sound Transmission Test.

than when the sound source is not operating. If the background level in the receiving room approaches the level of the signal coming through the partition, the results will be highly questionable.

   d. The numerical difference between the two levels will be the measured Noise Reduction, assuming little or no flanking past or around the partition.

     Significant "leaks" can usually be heard audibly or detected by a sharp rise in SPL as the microphone is moved toward the joints at the floor, ceiling, or walls or toward any penetration of the partitions by pipes, electrical services, etc.

   e. In carpeted, furnished rooms, the Noise Reduction will be approximately equal to the STC rating (Sound Transmission Class) of the partition.

     In bare, unfurnished rooms, the STC value will be approximately 3 to 5 points *higher* than the measured Noise Reduction.

     This same method is applicable to measuring the performance of floor/ceiling assemblies.

2. The Field Impact Sound Transmission Test procedure is essentially the same as that outlined under "Laboratory Test."

3. Measurement of the absorptivity of materials or surfaces in a completed room is a highly questionable procedure, in most cases. The *total* absorptivity of the room can be determined by measuring the reverberation time within the room and deducing the absorption from the Sabine formula. (See Section II, page 57.)

In a room completed (but unfurnished) except for the ceiling, with a relatively hard, nonabsorbent exposed lower surface of the floor above, measurements of reverberation time can be made; and then measurements can be repeated after the ceiling is installed. The added absorption can be calculated from the change in reverberation time, and the absorption coefficients of the ceiling can be determined by dividing total *added* absorption units by the ceiling area. The results, however, may vary widely from laboratory tests of the same material, since laboratories go to great lengths to produce a truly "diffuse" sound field (a condition rarely occurring in the field), and the addition of the ceiling may reduce room volume significantly if the ceiling is suspended.

Reverberation decay tests are frequently conducted in the field to determine whether a design goal has been met or whether absorption must be added or removed to provide the desired reverberance in the space. The procedure usually involves introducing a very loud, brief, impulsive sound and recording it on magnetic tape for later transfer to graph paper. The slope of the decay curve will give an indication of the reverberation time for a 60 dB decay—by definition, the reverberation time. An alternative procedure uses a Type I sound level meter, with its output fed directly to a graphic level recorder, which produces a logarithmic decay curve directly in "real time."

In general, though, it is best to rely on competent consultants when any reverberation or absorptivity tests are required. There are so many pitfalls that only the most experienced practitioner can interpret such field data in a meaningful way.

FIELD MEASUREMENTS

Complete and detailed procedures for many types of sound, noise, and vibration measurements are published by standards organizations such as ISO, ANSI, ASTM, ASME, SAE, and others. When such procedures are available, they should be followed as closely as possible. Likewise, many zoning and other ordinances and governmental regulations prescribe in detail how measurements must be made and the data reported.

Usually, for legal contests and similar situations where the expertise of those making the measurements must be established, it is advisable to use the services of experienced professionals.

Often, though, some simple measurements, employing simple, standard instrumentation, can be useful to professionals not engaged full-time in acoustical work.

Some simple recommendations for such measurements follow: (A wealth of excellent information regarding sound, noise, and vibration measurements is available; refer to the "Bibliography," Section III.)

1. Use as little equipment as possible, and choose the best quality equipment the budget will permit and the job's requirements will justify.
2. Read the user's instructions!
3. Calibrate all equipment carefully and frequently during *every* job, and *always* at the beginning and end of every job.
4. Whenever possible, monitor all measurements requiring simultaneous recording of data. Be sure the meter readings are consistent with what the tapes (magnetic or paper) indicate. Record all instrument settings, preferably on a data sheet *and* audibly on magnetic tape or with pen on paper tapes or graphs.
5. Beware of 10 dB mistakes! The decade attenuator of most equipment has been the downfall of many measurements.
6. Check measurements against what you know you should be finding; for example, in a quiet room, the meter should read about 65 dBA when you talk to it in a normal voice at 36 in. (1 m) from the microphone; if conversation in the area is almost impossible, you should be reading levels over 85 dBA.
7. Be sure to notice and record whether you are on the A, B, C, or linear network of the meter. A mistake here could invalidate all measurements, particularly if you aren't certain which network you were using.

8. Locate the microphone where it is measuring the sound field you really want to measure—not the region of pressure doubling next to a wall or in the near field of an adjacent sound source which may be overpowering the source you want to measure—and orient it properly.

9. Attach the right type of accelerometer properly to the proper location on the vibrating body you are studying. A huge accelerometer which loads a thin, light panel, vibrating at high frequencies might totally change the panel's vibration.

10. When octave-band (or narrower bands) data are required, be sure to record the *overall* level, too, on the linear network; and then add up the octave-band numbers to see if the total energy equals the overall level measurement. If it doesn't, find out why.

11. If you have octave-band data, and you forgot to get (or didn't think you'd need) A-weighted data, you can easily calculate it, using the following table:

TABLE 3-11e   Difference Between Linear and A-weighted Level

Octave-band Center Frequency (cps or Hz)	Difference from Flat or Linear Level (dB)
31.5	−39.4
63	−26.2
125	−16.1
250	− 8.6
500	− 3.2
1000	0
2000	+ 1.2
4000	+ 1.0
8000	− 1.1

12. Be sure to use the *slow* needle when it is called for; but remember—some measurements require the *fast* needle, the *impulse* setting, or the *impact* setting. Use the right setting, check it frequently, and record it. The wrong setting might make an enormous difference; in fact, it usually does!

13. And finally, be *sure*—absolutely *sure*—you have measured the background—when your observed source is *not* operating. If the background level is too close to the measured source level, you may have questionable or even useless data.

The reader will notice that we have listed 13 caveats—not because we are superstitious, but because Murphy's law ("if anything can go wrong, it will") was developed in acoustical measuring procedures. So try to eliminate at least the more common errors, and hope you don't discover others when you are back at the office, a thousand or more miles from where the measurements were made.

Trouble-Shooting

In existing buildings, identifying acoustical problems, determining their causes, and deciding upon the most economical solutions are frequently the tasks of the professional. It is remarkable how many thousands of dollars are spent "treating" the wrong part of the building or prescribing the wrong "treatment" for the problem.

Identifying the problem consists of observing and measuring in the spaces involved. A subjective description will usually give clues to the objective cause of the problem; but don't accept the "diagnosis" of the occupants, and don't jump to conclusions. The immediate, superficial evaluation may disguise the real problem.

Associating the effect with the cause is not always simple. When the source of the noise is obvious—such as a single noisy machine located in an otherwise quiet area—it is easy to relate cause and effect. (For example, see Table 3-10, page 154.) For the musically trained, the pitch of the sound can be identified simply by listening, and the frequencies involved can be determined. Even more direct is the simple expedient of turning off (or on) all equipment, one piece at a time, until the offender is found. Occasionally, special measuring equipment and techniques are required.

In subsequent tables, correlation between subjective complaints, possible causes of the problems, and likely solutions is shown; simple procedures for investigation of the more common problems are discussed; and available sound control options are listed.

TABLE 3-12 Typical Acoustical Problems

Problem	Possible Causes	Solution
It's so "noisy," I can't hear myself think	High levels Excessive reverberation	Absorption
	Excessive transmission	Sound isolation
	Excessive vibration	Vibration isolation
	Focusing effects	Eliminate cause of focusing effects

TABLE 3-12 (continued)

It's "not loud enough" at the rear of the room	Room too large	Electronic amplification
	Improper shape	Alter room shape
	Lack of reflecting surfaces	Add reflecting surfaces
	Poor distribution Too much absorption	Eliminate absorption on surfaces needed for reflection
There are "dead spots" in the room	Poor distribution	Reflecting surfaces
	Improper shape	Eliminate cause of focusing effects
	Echoes	Alter room shape
Sound "comes right through the walls" of these offices	Sound leaks	Eliminate leaks
	Sound transmission	Sound isolation
	Vibration	Vibration isolation
	Receiving room too quiet	Masking
	Poor room location	Proper room layout
I can hear "machine noises" and "people walking around upstairs"	Vibration	Vibration isolation
	Sound transmission	Sound isolation
	Poor room location	Proper room layout
"Outside noises" drive me "nuts"	Poor room location	Proper acoustical environment
	Sound leaks	Eliminate leaks
	Sound transmission	Sound isolation

TABLE 3-12 (continued)

Problem	Possible Causes	Solution
Speech and music are "fuzzy, indistinct"	Excessive reverberation	Absorption
"Little sounds" are most "distracting" in here	Background level too low	Masking
	Room too "dead"	Optimum reverberation
There's an annoying "echo"	Echo Flutter	Proper room shape
	Focusing effects	Eliminate cause of focusing effects
	Excessive reverberation	Absorption
I can hear everything the fellow across the office says	Room too "dead"	Optimum reverberation
	Background level too low	Masking
	Focusing or reflection	Eliminate focusing or reflection
It doesn't sound "natural" in here	Flutter	Alter room shape Add absorption
	Distortion caused by improper sound system	Proper sound system
	Selective absorption Room too "dead" — low reverberation time	Proper type and amount of absorption
It feels "oppressive" in here	Reverberation time too low	Proper amount of absorption
	Background sound level too low	Use of background and masking sound

ECHO, FLUTTER, REVERBERATION, AND FOCUSING EFFECTS

Clap your hands sharply in the room; or use a loud, short burst of a tone; or even a loud shout.

Echo

Is there a single, clear, distinct echo? If so, look for a large, flat, distant surface that should have been tilted, broken up with splays, or set at an angle. If you can't make these corrections, consider applying acoustical absorbents.

Flutter

Is there a "rattle," or "clicking, buzzing, clattering" sound, or a "throbbing" sound which can be described as a "flutter"? If so, look for parallel walls sufficiently far apart or sufficiently long (such as in a corridor) to permit multiple echoes. This "shooting gallery" effect often occurs after acoustical tile has been applied to the ceiling of the room. It was lost in the reverberation before, but now it is clear and distinct. If you can't break up, tilt, or "angle" one of the walls, put enough acoustical absorbent on one of them to stop the flutter.

Reverberation

Does the sound "persist" and slowly die away to inaudibility? Does speech get jumbled, confused? Is it hard to separate and distinguish individual syllables? This high-speed, multiple echo effect means that sound absorption must be introduced into the space.

Focusing Effects

Does the sound level seem to be very high in only one area? Is there a "whispering gallery" effect so that a fairly quiet sound can be heard distinctly in a particular area quite distant form the source? If so, look for concave-curved surfaces. Break them up or load them with acoustical tile.

TRANSMISSION

A portable sound source of fairly high and steady level is necessary. A common household vacuum cleaner is excellent.

Leaks

Chances are almost ten-to-one that holes, leaks, and openings are the most serious problem. Put the vacuum cleaner in the "offending room" and turn it on. Then go into the "complaining" room.

"Follow your ears." Does the sound get louder as you approach the door? The windows? A ventilating grille? The convector under the window? The big recessed light fixture in the ceiling?

If so, you can be sure that a hole or leak or "speaking tube" exists. Close the leaks; or put duct lining or baffles in the ducts. If the electrical outlet boxes or medicine cabinets are back to back on the wall, little can be done to close these leaks.

Wall or Ceiling Transmission

If there are no leaks, does the sound get louder as you approach the common wall between rooms? Put your ear against the wall. Does the sound seem to "come right through"? If so, you'll have to make the wall heavier. Laminate gypsum board to both sides; or put up a separate wall, not attached to the offending wall.

If the wall is not the offender, how about the ceiling? Is it just suspended acoustical tile with a plenum common to all other rooms? If so, forget the wall, it's probably far better than the ceiling.

If the ceiling is the offender, install vertical baffles from ceiling to slab above the partitions. Or, if possible, lay gypsum board above the entire suspended ceiling.

Structure-borne Sound

Do you hear footfalls, equipment noises, or bumps coming from upstairs? Have someone walk around upstairs, tapping the floor with a cane or hard stick. Put your ear to the wall. Can you hear the sound distinctly? If so, have a carpet or resilient flooring material installed upstairs.

If you can't particularly hear the sound at the wall, but it is still easily heard all over the room, the carpet will still work. Or a heavy, resiliently hung ceiling in your space will help.

Above all, in any transmission problem, don't waste your time or money sticking up or hanging acoustical tile.

VIBRATION

Your ears or finger tips are the best measuring instruments for investigating this problem.

Place your ear tightly against the wall or ceiling; or place your finger tips lightly against the surface in question. Can you "feel" or hear the vibration? Is it the kind of steady vibration that obviously comes from some mechanical equipment or a large electrical transformer?

Search the entire building for possible causes. When you find them, isolate them *and all their attachments, pipes, and appurtenances from the structure.*

Source–Path–Receiver

Sound originates with a SOURCE, travels via a PATH, to a RECEIVER. Sound Control consists of modifying or "treating" any or all of these three elements in some manner.

SOURCE: The most effective control measures often involve eliminating noise at the source, *or* in reinforcing the source when necessary. For example:

1. Balancing moving parts, lubricating bearings, improving aerodynamics of duct systems, etc.
2. Modifying parts or processes.
3. Changing to a different, less noisy process.
4. Amplifying or reinforcing the source.

PATH: The most common sound control measures usually involve acoustical "treatment" to absorb sound; but equally important is the use of barriers to prevent airborne transmission of sound; interrupting the path with carefully designed discontinuities; and the use of damping materials to minimize radiation from surfaces. Another very important approach involves reinforcing the direct sound with controlled reflections from properly designed reflective surfaces.

RECEIVER: Often the simplest, most effective sound control involves protecting the receiver; enclosing him within adequate barriers; or equipping him with personal protection devices (earplugs or muffs), rather than trying to enclose or modify huge sources or an entire room or building.

All of these principles are illustrated with examples and suggestions in preceding pages of this section. We refer the reader again to Section II for a discussion of the principles involved.

TABLE 3-13 Available Options in Noise Control

Objectives of Sound Control Efforts	Quiet the Source (2)	Barriers or Enclosures	Vibration Isolation or Damping	Absorption	Masking	Personal Protection
Reduce the general noise level to:						
Improve communication	x		x	x*		
Increase comfort	x		x	x*		
Reduce risk of hearing damage	x		x	x*		
Reduce extraneous, intruding noise to:						
Increase privacy		x*	x		x	
Increase comfort		x*	x			
Improve communication		x*	x			
Protect many people against localized source producing damaging levels	x	x*	x	x		x
Protect many people against many distributed sources producing damaging levels	x	x	x	x		x*
Protect one person against localized source producing damaging levels	x	x(3)	x			x*
Protect few persons against many distributed sources producing damaging levels		x(3)				x*
Eliminate echoes and flutter				x(1)		
Reduce reverberation				x*		
Eliminate annoying vibration			x*			

Notes: *indicates the most likely solution(s)
1 assumes that the configuration of the reflecting surface(s) cannot be modified
2 always the best and simplest means of eliminating noise, if practical and economical
3 a closed booth or small room for the person(s) is often feasible
No mention has been made of another option — reinforcement — since its purpose is to increase
levels. It is the only available option when the signal level must be increased.

Industrial Noise Control—Case Histories

GENERAL

While the same principles of noise and vibration control apply to industrial noise problems as to those associated with architectural and mechanical equipment problems, actual execution of the solutions is normally quite different.

"Feasible" engineering methods of worker exposure control in industrial plants usually involve limited, "custom" solutions to specific, individual problems. The word "feasible" as used in this book means:

1. The solution is possible. The technology exists, and no scientific or technological "breakthroughs" are required.
2. The solution is practical. It can be applied to the particular situation without disrupting the operation, deteriorating the quality of the work, or making impossible the human supervision or control necessary.
   This *may* involve process modification, equipment changes, or even automation.
3. The solution is economic. It can be used without such cost penalties that the operation would become uneconomic (*not* simply that the solution might increase costs to some degree).

COST/BENEFIT ANALYSIS

It is always wise, before attempting to choose a noise control method in industry, to weigh the potential costs against the possible benefits of the procedure. This usually means asking:

1. How many workers are involved?
2. Are workers concentrated in a small area or widely distributed throughout the plant?
3. Are the noise sources concentrated or distributed?
4. Which approach will protect the most workers?
5. Is it simpler to isolate the noise sources or the operators?
6. Which approach will provide the greatest noise reduction at the lowest cost—that is, how many dollars per dB are involved? How many dollars per worker?
7. Can the option chosen actually provide the reduction required?
8. What productivity penalty will result?

Table 3-14, Section III, page 214, "Industrial Noise Control Cost/Benefit Analysis", is useful in answering these questions.

TABLE 3-14   Industrial Noise Control Cost/Benefit Analysis

Control Option	Attenuation dBA	Unit Cost per sq ft or each	Productivity Penalty
Absorption	3 to 5	$.50 to $2.00	None
Damping	3 to 10	$.20 to $4.00	None
Barriers	5 to 15	$3.00 to $6.00	Up to 15%
"Glove Box" Booths	3 to 15	$250 to $500 each	Up to 20%
Machine Enclosures	5 to 50	$5.00 to $20.00	Up to 25%
Worker Shelters	5 to 25	$250 to $3500 each	None

These are feasible ranges of performance; typical cost ranges; and productivity penalty data based upon actual, on-the-job case studies in the U.S.A.

MOCK-UPS

Since custom solutions are often required for industrial noise problems, it is advisable, whenever possible, to mock up the proposed enclosure, shield, or barrier in some cheap, easily worked material—corrugated paper, plywood, or hardboard. Cutting, fitting, and adjusting can be done while the machine is in operation, and the operator's comments and suggestions can be used effectively in the design.

This study might show, for example, that fixed, rigid panels are unacceptable; flexible, readily moved "curtains" or similar devices may be required. It might also show that so little attenuation will be provided that another approach will be necessary.

CASE HISTORIES

The following case histories, taken from actual field installations, are representative of feasible industrial noise control approaches. Each case is unique, and each case involved the calculations and procedures described above. It is not advisable to assume that any of the solutions can be applied directly and without modification to another situation. However, each case indicates the general approach to common problems and the costs and results expected.

It is imperative to remember that requirements for even ostensibly identical machines vary according to location, use, and other specific needs of the particular job. It is not safe to extrapolate directly from one successful solution to a new situation without investigating carefully all aspects of the job. Access to machines, ventilation requirements and limits to permissible heat build-up, durability of housings or enclosures, and a host of specific details must be investigated before making the installation, or it may prove to be a costly failure in spite of its acoustical success.

Case History No. 1

A continuous, line source, affecting a large area.

The Problem

Screw and bucket conveyors throughout the production area of a chemical plant exposed many workers to levels reaching 110 dBA. Frequent and ready access to the conveyors for maintenance and clearing jams was essential.

The Solution

The conveyors were wrapped with a composite of urethane foam and leaded vinyl, with a snap-on fastening arrangement to facilitate access to any point in any line. (See Figure 3-CH-1.)

The Result

Noise levels reductions of almost 30 dBA were effected, and the conveyors were eliminated as significant noise sources in the plant.

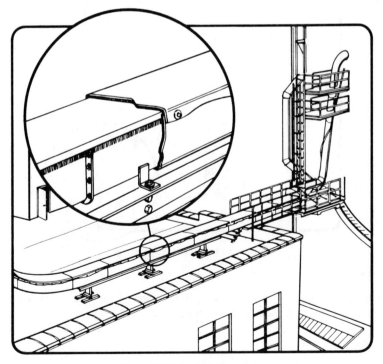

Figure 3-CH-1 Conveyor lines.

Case History No. 2

A single, concentrated, steady source, affecting the surrounding community.

The Problem

A centrifugal pump, located out-of-doors, adjacent to an electroplating plant, operated continuously, 24 hours a day. It produced levels at the plant property line almost 20 dBA in excess of the community ordinance limits.

The Solution

The pump was enclosed in a shroud made of a composite of urethane foam and leaded vinyl supported by a galvanized structural steel frame. Cooling air circulated along the bottom of the enclosure, and ready access to the top was provided by means of a snap-off cover. (See Figure 3-CH-2.)

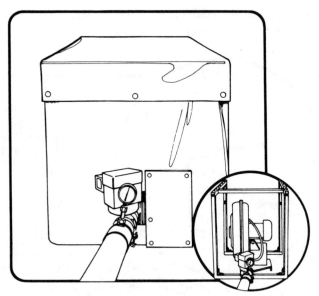

Figure 3-CH-2 Centrifugal pump.

The Result

A noise level reduction of 23 dBA at the property line resulted, satisfying the community ordinance requirements with a margin of safety.

Case History No. 3

A large, steady source, affecting workers in the adjacent area.

The Problem

A large, hydraulic power unit for several machines in a metal working factory, operating continuously, exposed adjacent workers to levels in excess of 103 dBA.

Free circulation of air for cooling and to carry off fumes was imperative; and frequent access to the machine for adjustment and inspection was necessary.

The machine sprayed a fine mist of oil and solvent, precluding the use of porous absorbents near the machine.

The Solution

A flexible vinyl "curtain" enclosure with a Velcro closure strip was placed around the installation. Hanging, absorptive baffles with impervious, heat-sealed wrapping were located at the top of the enclosure. (See Figure 3-CH-3.)

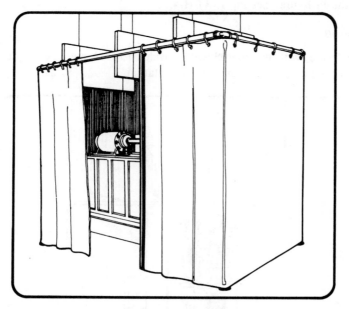

Figure 3-CH-3 Hydraulic power unit.

The Result

A noise level reduction of 18 dBA was effected in the adjacent area, eliminating the machine as a significant noise source for the adjacent workers. Levels within the enclosure were not reduced, of course, but worker exposure within the enclosure is limited to brief periods for adjustment and inspection.

Case History No. 4

A single, high-speed, continuous impact source, affecting workers in a large area.

The Problem

A small, high-speed, automatic stamping press, operated almost continuously throughout the day, producing noise levels of 95 to 100 dBA in the surrounding area.

The operator's duties involved only loading the strip stock, changing dies, removing finished parts, and monitoring the operation visually.

The Solution

A flexible, composite "curtain" of leaded vinyl with a coated urethane foam layer on the inside was hung from a track around the entire press. A sliding panel of the same material with a transparent vinyl viewing panel provided access to the inside of the enclosure. Hanging, absorptive baffles were located above the top of the enclosure. (See Figure 3-CH-4.)

The Result

The enclosure, with the door closed, provided 15 dBA noise level reduction in the adjacent area, including the operator's station.

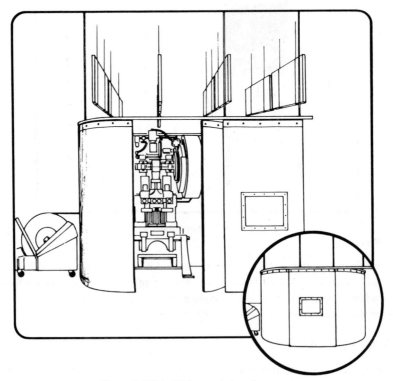

Figure 3-CH-4 High-speed stamping press.

Case History No. 5

Multiple, high-speed, continuous impact sources, affecting workers in a large area.

The Problem

Several small, high-speed, automatic stamping presses in a row, operated almost continuously, producing noise levels of 95 to 100 dBA in the surrounding area.

Operators' duties included loading the strip stock, changing dies, and removing finished parts.

Oil spray was a serious problem, and fire hazard was high enough to limit enclosing materials to fire-resistant and nonabsorbent products. Spacing between presses was minimal, and access limited.

The Solution

Encircling "curtains" of leaded vinyl, suspended from frames above each press, were made so that they could be hoisted up individually, out of the way for access to the machines. Hanging, absorptive baffles were located above the row of presses. (See Figure 3-CH-5.)

Figure 3-CH-5 High-speed stamping presses.

The Result

Noise level reductions of 7 to 8 dBA in the working area resulted from the use of the curtains.

Case History No. 6

A large, single source, operating continuously for long periods, affecting workers in the adjacent areas.

The Problem

A Raymond Mill pulverizer in a food processing plant produced noise levels of 103 dBA.

Operators monitored the process intermittently. Adjacent workers were exposed continuously whenever the mill operated.

Coatings on the pulverizing drum were not permitted and any materials used near the process had to meet strict health and sanitation standards.

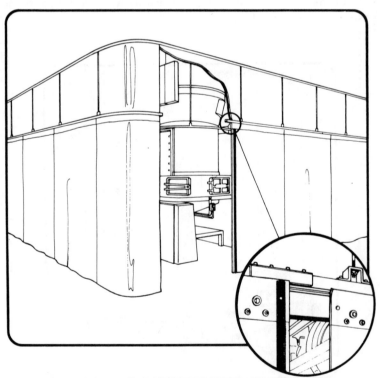

Figure 3-CH-6 Pulverizer.

The Solution

A two-tier vinyl curtain arrangement was used, with the upper tier made of a composite of loaded vinyl (non-lead loading) faced with a sealed layer of urethane foam. The lower tier consisted of a "curtain" of the same loaded vinyl, with several operable panels for ready access for large hand trucks and similar equipment. Completely sealed, absorptive baffles were hung above the entire area, within the upper tier of the enclosure. (See Figure 3-CH-6.)

The Results

The enclosure, fully closed, provided 15 dBA reduction in noise level in the adjacent areas.

Case History No. 7

An extensive test operation, with many continuously operating sources, affecting workers in adjacent areas.

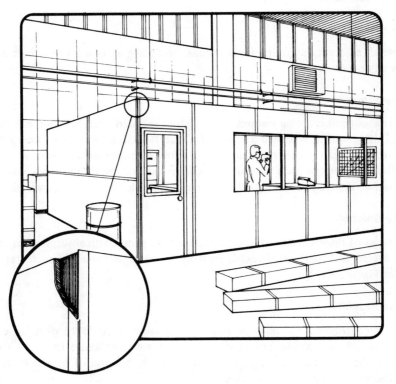

Figure 3-CH-7 Run-up room for chain saws.

The Problem

The final production step in the manufacture of chain saws involved operating them at idle for several minutes, with alternating periods of higher speeds, and making necessary adjustments. As many as 35 saws might be operating simultaneously. Levels up to 115 dBA inside the run-up area were common.

The Solution

A total enclosure was constructed of 4-in. thick steel panels, with the inside surface perforated and covering a 4-in. layer of mineral wool wrapped in a thin plastic film. The enclosure was divided into two rooms, one for the brief adjustments and the other for the continuously operating saws. (See Figure 3-CH-7.)

The Results

The noise level outside the room was reduced to below 85 dBA. Inside the room, where adjustments were made, operator exposure was reduced to permissible OSHA schedules.

Case History No. 8

Several, adjacent, intermittent operations, producing an unacceptably high level for all operators.

The Problem

A row of hand chipping stations in a foundry generated levels reaching as high as 115 dBA, with a weighted average of about 100 dBA. Operations were intermittent, and rarely were more than two stations operating simultaneously.

Ready access to each station for fork trucks and overhead cranes was imperative. Ventilation and lighting requirements precluded enclosing each station.

The Solution

A series of three-sided booths, fabricated of steel panels with absorptive material protected with perforated metal, partially enclosed and shielded each station. (See Figure 3-CH-8.)

The Results

Noise contributed by adjacent stations was reduced 10 to 12 dBA for each operator. Levels within each operator's booth were reduced 5 to 7 dBA from those the operator formerly generated (probably because of reduction in reverberation in the chipping room).

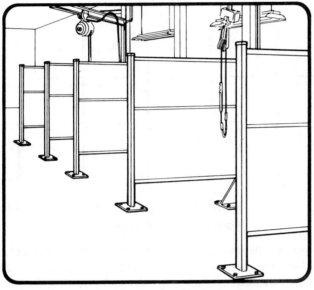

Figure 3-CH-8   Foundry chipping booths.

Case History No. 9

An extended assembly line with multiple, intermittent noise sources.

The Problem

An extended assembly line for spot-welding light-gauge sheet metal panels into a finished cabinet, stationed several operators, side-by-side, along a moving belt. Each operator performed his work on each piece in a few seconds and passed the part on, along the belt, to adjacent operators, who performed similar tasks. As the part progressed down the line, it was subjected to frequent impacts as well as to the welding operation. The general level in the area was approximately 94 dBA, but individual operators were exposed to 96 to 97 dBA for brief periods during the operation of their own welders.

Ready access to each station to deliver and remove parts and bins with fork lift trucks was imperative. Lighting and ventilation could not be interfered with. Production engineers would not accept any rigid, permanently attached panels or barriers. Flying sparks caused a continual fire hazard.

The Solution

Time studies showed that each operator spent less than 19% of his $7\frac{1}{2}$-hour working day actually operating his own welder; the balance of the time was used

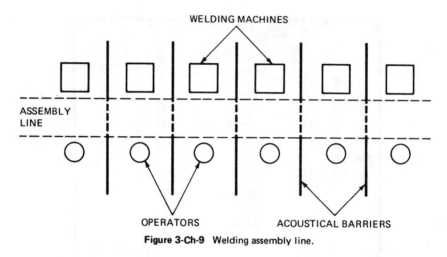

Figure 3-Ch-9   Welding assembly line.

to put the part on the assembly line, pick up a new part, and wait for the next part on the line. If he could be protected from the noise of the adjacent stations, he could tolerate his own operation's exposure under the OSHA schedule.

High, flexible leaded vinyl curtains, covered on both sides with quilted Fiberglas and woven glass-fiber cloth, were erected between stations. Transparent, flexible vinyl strips were inserted into the curtains over the assembly line to permit parts to pass through, and vision panels of the same material were located in each curtain to permit eye contact between stations. (See Figure 3-CH-9.)

Sheet metal surfaces were damped and all impact points cushioned with tough elastomers.

The Results

Except during the brief period when each operator activated his own welder, levels in each station fell to 87 to 88 dBA. During operation of his own welder, each operator was still exposed to approximately 96 dBA. However, this level is permissible for more than 3 hours a day under OSHA schedules. The combination of "engineering and administrative" controls resulted in acceptable worker exposure.

Case History No. 10

An extended line of continuously operating, identical sources, affecting all operators and workers in the plant.

The Problem

Two rows of automatic screw machines (48 total) in a low, long, narrow building, were tended by 24 operators, one for each pair of machines.

Operating on octagonal stock, the machines produced a steady level of approximately 106 dBA throughout the plant.

The Solution

Since a new building of almost identical dimensions was being planned for the machines, management elected to undertake a study program to reduce operator exposure to acceptable levels in the new plant.

A series of noise reduction measures was programmed to permit continued operation while attempting to lower noise levels to acceptable limits when the new building was completed and occupied. The steps included:

1. Elastomer stock tube liners;
2. Absorptive wall and roof surfaces in the new plant;
3. Operator "screens" or barriers.

(See Figure 3-CH-10.)

Wall and roof surfaces were prefabricated, perforated metal liner panels, with thick Fiberglas insulation batts between surfaces. Operator "screens" consisted of three-sided hemi-hexagonal, free-standing screens, made of 4-in thick panels, solid sheet metal exterior, perforated metal interior, covering dense mineral wool batts, enclosed in thin plastic films. Panels were 96 in high (2.44 m) and 36 in wide (1 m), with small viewing windows in each panel, glazed with laminated safety glass. Operators located their tool chests and chairs behind the screens, near the building wall.

The Results

Figure 3-CH-10b shows, graphically, the results of each successive step in the program. Levels behind the barriers, at the operator stations, fell to 87 dBA.

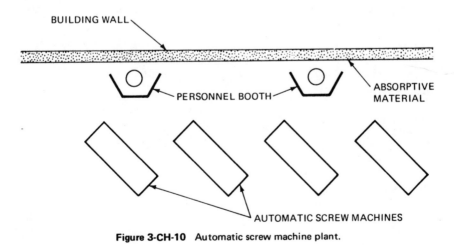

BUILDING WALL

PERSONNEL BOOTH

ABSORPTIVE MATERIAL

AUTOMATIC SCREW MACHINES

Figure 3-CH-10 Automatic screw machine plant.

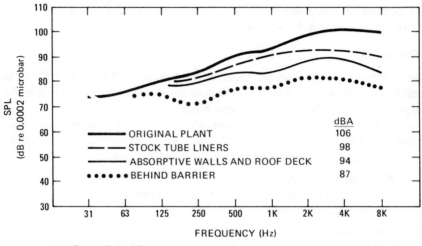

Figure 3-CH-10b Noise levels: automatic screw machine plant.

Operators were still required to spend about 20% of their operating day outside the barriers, in the 94 dBA levels, but this is acceptable exposure under OSHA schedules.

Glossary

Insofar as practicable, the definitions contained in this glossary conform to the spirit of the definitions of terms according to ANSI, ASTM, ISO, and other recognized standards organizations. Whenever the "official" definitions are complex or obscured by technical verbiage, we have used the common sense language of everyday usage and the idiom of the practicing architect or engineer.

Absorbents, Diaphragmatic Materials which flex under sound pressures and vibrate as a diaphragm, dissipating acoustic energy within their structure as heat and as mechanical energy of vibration.

Absorbents, Sound Materials which absorb sound readily; usually building materials designed specifically for the purpose of absorbing acoustic energy.

Absorption, Sound Conversion of acoustic energy to heat or another form of energy within the structure of sound-absorbing materials.

Absorption Coefficients, Sound The ratio of sound-absorbing effectiveness (at a specific frequency) of an area of acoustical absorbent to the identical area of perfectly absorptive material; usually expressed as a decimal value (such as 0.70) or in per cent.

Acoustic An adjective used in conjunction with a basic property of sound, such as "acoustic energy."

Acoustic Energy The total energy of a given part of the transmitting medium minus the energy which would exist in the same part of the medium with no sound waves present. The energy added as the result of the sound vibrations.

Acoustical An adjective used in conjunction with apparatus, design, or other nouns or verbs connected with sound control, such as "acoustical tile," "acoustical analysis."

Acoustical Environment All of the factors, interior or exterior, which affect the acoustic conditions of the location, space, or structure under consideration.

Acoustical Tile Acoustical absorbents produced in the form of sheets or units resembling tiles; usually 12 X 12 in. (30.5 X 30.5 cm) or multiples thereof.

Acoustical Treatment The use of acoustical absorbents, acoustical isolation, or any changes or additions to the structure to correct acoustical faults or improve the acoustical environment.

Acoustics The science of sound, including its production, transmission, and effects.

Acoustics, Architectural The acoustics of buildings and structures.

Airborne Sound Sound transmitted through air as a medium rather than through solids or the structure of a building.

Amplification, Electronic Increasing the intensity level of a sound signal by means of electrical amplification apparatus.

Amplitude The maximum displacement to either side of the normal or "rest" position of the molecules, atoms, or particles of the medium transmitting the vibration.

Analysis, Acoustical A detailed study of the use of the structure, the location and orientation of its spaces, and a determination of noise sources and the desirable acoustical environment in each usable area.

Analyzer, Frequency Electrical apparatus capable of measuring the acoustic energy present in various frequency bands of a complex sound.

Attenuation, Sound Reducing the intensity of a sound signal.

Audible Capable of producing the sensation of hearing.

Audibility, Threshold of The sound pressure at which average normal hearing apparatus begins to respond. (Usually 0.0002 microbar or 20 μPa.)

Background level The normal sound level present in the space above which speech, music, or similar specific wanted sound must be presented.

Coincidence The condition (or frequency range) at which the velocity of the parallel component of the sound wave incident upon a panel equals the velocity of the shear wave (*which see*) in the panel.

Communication The signals or stimuli (or their transmission) which produce reactions, orient us in our environment, and furnish information on which to base decisions.

Compliance Roughly, the ease with which a panel of a material can be flexed by application of a force or pressure.

Coupling Any means of joining separated masses of any media so that sound energy is transmitted between them.

Cycle The entire sequence of movement of a particle (during periodic motion) from rest to one extreme of displacement, back through rest position to the opposite extreme of displacement, and back to rest position.

Damping Any means of dissipating or attenuating vibrational energy within a vibrating medium. Usually the energy is converted to heat.

Decibel A division of a uniform scale based upon 10 times the $\log_{10}$ of the relative intensity of sound intensities being compared.

Diffraction Roughly, the ability of a sound wave to "flow" around an obstruction or through openings with little loss of energy.

Diffusion Dispersion of sound within a space so that there is uniform energy density throughout the space.

Discrimination The ability of the hearing apparatus to discern discrete, particular signals in a complex sound field.

Dispersion The scattering or distribution of sound in a space. (*See also* Diffusion.)

Distortion Any change in the transmitted sound which alters the character of the energy-frequency distribution within the signal so that the sound being received is not a faithful replica of the source sound.

Distribution The pattern of sound intensity levels within a space; also, the patterns of sound dispersion as the sound travels within the space.

Echo Any reflected sound which is loud enough and received late enough to be heard as distinct from the source.

Elastic medium Any substance in which strain or deformation is directly proportional to stress or loading.

Fidelity The faithful reproduction of the source sound.

Flanking paths Transmission paths which transmit acoustic energy around a sound barrier; paths which "by-pass" the intended barrier.

Flutter A rapid reflection or echo pattern between parallel walls, with sufficient time between each reflection to cause a listener to be aware of separate, discrete signals.

Focusing Concentration of acoustic energy within a limited location in a room as the result of reflections from concave surfaces.

Frequency The number of complete cycles per second of a vibration (or other periodic motion). Usually stated in cycles per second (cps) or Hertz (Hz).

Frequency, Infrasonic Frequencies below those by which human hearing apparatus is stimulated; usually under 16 cps (Hz).

Frequency, Ultrasonic Frequencies above those by which human hearing apparatus is stimulated; usually above 18,000 to 20,000 cps (Hz).

Hearing The subjective response to sound, including the entire mechanism of the internal, middle, and external ear; and the nervous and cerebral operations which translate the physical operations into meaningful signals.

Hearing loss An increase in the threshold of audibility, at specific frequencies, as the result of normal aging, disease, or injury to the hearing organs.

Impact The sharp, rapid contact between two solid bodies.

Impact Noise Reduction A single-number rating system which compares the impact isolation of a test specimen with a standard contour. (Also given in terms of a comparable "Impact Insulation Class." Strictly, only when tested in accordance with a specific test procedure.)

Impedance A complex ratio related to the sound absorption or transmission characteristics of acoustical materials. Similar to electrical or mechanical impedance. Actually, the rate at which a given volume of any material can accept energy.

Inertia The tendency of a mass to resist any change in its state of motion or rest.

Intensity The rate of sound energy transmitted in a specified direction through a unit area.

Intensity level Ten times the $\log_{10}$ of the ratio of intensity of the sound to a reference intensity.

Isolation, sound Materials or constructions (or the use of such materials or constructions) which resist the passage of sound through them.

Leaks, sound Any opening which permits airborne sound transmission.

Level, Sound A measure of Sound Pressure Level as determined by electrical equipment meeting ANSI requirements. Unless specifically stated otherwise, levels refer to root-mean-square of sinusoidally varying level.

Level meter, sound An electrical instrument for determining Sound Pressure Level.

Limpness Roughly, inelastic motion; motion in which the material under stress does not recover from the strain upon removal of stress.

Longitudinal wave A wave in which displacement of the molecules in the medium is parallel with the direction of propagation of the wave.

Loudness The effect on the hearing apparatus of varying sound pressures and intensities.

Loudness Level The Sound Pressure Level in decibels (relative to 0.0002 microbar or 20 μPa) of a simple tone of 1000 cps (Hz) frequency.

Masking The increase in threshold of audibility of a sound necessary to permit its being heard in the presence of another sound.

Mass The quality of matter which permits it to resist acceleration; the quality of matter which produces the effect of inertia.

Mounting, resilient Any mounting, attachment system, or apparatus which permits room surfaces or machinery to vibrate without transmitting all of the energy of vibration to the structure.

Mounting, tile The method of attaching acoustical tile to the building structure or building surfaces.

Natural frequency The frequency at which a resiliently mounted mass would vibrate, when set into vibration, under the influence of gravity alone, with no added force or constraints. (Often called "resonant frequency.")

Noise Any unwanted sound.

Noise Criteria Refers to a set of contours, roughly corresponding to the ear's response to Sound Pressure Level at various frequencies, which define the background sound level existing within a space.

Noise reduction The reduction in level of unwanted sound by any of several means.

Noise Reduction Coefficient The arithmetic average of the sound absorption coefficients at 250, 500, 1000, and 2000 cps (Hz).

Noy A "unit" or term applied to the divisions of a scale, comparable with the loudness scale, of relative "annoyance" or "noisiness" or sound.

Pain, threshold of A Sound Pressure Level sufficiently high to produce the sensation of pain in the human ear (usually above 120 dB).

Phon A measure of loudness level (on a logarithmic scale) which compares the effect of a sound to the effect of a 1000 cps (Hz) tone of a given Sound Pressure Level.

Pitch The physical response to frequency. The subjective response of the hearing mechanism to changing frequency.

Power, sound The time rate of acoustic energy flux. Usually stated as "Sound Power Level," in dB re 10^{-12} W.

Pressure, acoustic The instantaneous pressure at a point as a result of the sound vibration minus the static pressure at that point. The change in pressure resulting from the sound vibration.

Pressure Level, Sound   A value equal to 20 times the $\log_{10}$ of the ratio of the pressure of a sound to the reference pressure.

Quality   Usually refers to the spectral distribution of acoustic energy in a given sound.

Reflecting surfaces   Room surfaces from which significant sound reflections occur; or special surfaces used particularly to direct sound throughout the space.

Reflection   The return from surfaces of sound energy not absorbed upon contact with the surfaces.

Resonance   The natural, sympathetic vibration of a volume of air or a panel of material at a particular frequency as the result of excitation by a sound of that particular frequency.

Reverberation   The persistence of sound within a space after the source has ceased.

Reverberation time   The time in seconds required for a sound to decay to inaudibility after the source ceases. (Strictly, the time in seconds for the sound level at a specific frequency to decay 60 dB.)

Reverberation time, optimum   An empirically determined reverberation time, varying directly with room volume, which produces hearing conditions considered "ideal" by an average listening audience.

Room location   The orientation of a room with respect to other rooms on a given floor; with respect to other surrounding spaces; or with respect to other outside factors, including spaces outside of the building under consideration.

Room shape   The configuration of enclosed space, resulting from the configuration, orientation, and arrangement of surfaces defining the space.

Room volume   The cubic feet of space enclosed by the room surfaces.

Sabin   A measure of sound absorption of a surface, equivalent to 1 ft^2 of a perfectly absorptive surface. (The metric Sabin is equivalent to 1 m^2.)

Shear wave   A wave motion in which movement of the media is at right angles to direction of propagation of the wave. Wave propagation is parallel with the surface of the panel in which the wave propagates.

Sone   A measure of loudness (on a linear scale) which compares the effect of a sound to the effect of a 1000 cps (Hz) tone of 40 dB Sound Pressure Level.

Sound   A vibration in an elastic medium; usually in the frequency range capable of producing the sensation of hearing.

Sound control   The application of the science of acoustics to the design of structures and equipment, to permit them to function properly and to create the proper environment for the activities intended.

Sound Transmission Class   A single-number rating system which compares the Sound Transmission Loss of a test specimen with a standard contour.

Sound Transmission Loss   The ratio of sound energy incident upon a panel to the sound energy radiated from the opposite side (strictly, when tested according to ASTM E90-75).

Structure-borne sound Sound energy transmitted through the solid media of the building structure.

Transmission The propagation of a vibration through various media.

Transmission loss The decrease in power during transmission from one point to another (or through a panel, wall, etc.).

Unwanted sound Noise; interfering sound, whatever its source or nature.

Velocity The time rate of change of position of a reference point moving in a straight line.

Vibration An alternation in pressure or direction of motion.

Vibration isolation Any of several means of preventing transmission of sound vibrations from a vibrating body to the structure in which or on which it is mounted.

Wanted sound The audible signals which communicate necessary and desirable information or stimuli to the listener. (*See* Communication.)

Wave, sound A disturbance which is propagated in a medium in such a manner that at any point in the medium the displacement is a function of the time.

Wavefront A continuous surface which is a locus of points having the same phase at a given instant.

Wavelength The distance between adjacent regions where identical conditions of particle displacement, pressure, etc., occur.

Bibliography and References

General Acoustics: Beranek, "Acoustics," New York, McGraw-Hill, 1954. (Slightly "dated" but one of the "classics" in general acoustics.)

Kinsler and Frey, "Fundamentals of Acoustics," New York, Wiley, 1962. (A good, basic text.)

Stephens and Bate, "Acoustics and Vibrational Physics," Second edition, London, Arnold, 1966. (The newest, most comprehensive encyclopedic reference available.)

Acoustical Measurements: Peterson and Gross, "Handbook of Noise Measurement," Concord, Mass., General Radio Company, 1974. (Without question, the simplest, most complete and up-to-date book available; revised frequently.)

Broch, "Acoustic Noise Measurements," Naerum, Denmark, Bruel & Kjaer Company, 1973. (Very useful.)

Broch, "Mechanical Vibration and Shock Measurements," Naerum, Denmark, Bruel & Kjaer Company, 1976. (Very useful.)

Technical Review, Denmark, Bruel & Kjaer Company. (Published quarterly, with articles by numerous authors on various subjects. Very useful.)

Noise Control: Harris, "Handbook of Noise Control," New York, McGraw-Hill, 1957. (The major and most complete reference available.)

Beranek, "Noise and Vibration Control," New York, McGraw-Hill, 1971. (A curious mixture of theory and practice, but very useful.)

Olishifski and Harford, "Industrial Noise and Hearing Conservation," Chicago, National Safety Council, 1975. (A "smorgasbord" of useful information for industrial noise control; uneven but very helpful.)

Berendt and Corliss, "Quieting, a Practical Guide to Noise Control," Washington, D.C., National Bureau of Standards, 1976. (A very simple "cookbook" of useful tips, particularly for housing construction.)

Architectural Acoustics: Knudsen and Harris, "Acoustical Designing in Architecture," New York, Wiley, 1950. (Quite "dated," but one of the only books available on general architectural acoustics.)

Beranek, "Music, Acoustics, and Architecture," New York, Wiley, 1962. (A superb study of concert halls and large music performing spaces.)

233

Doelle, "Acoustics in Architectural Design," Ottawa, National Research Council, 1965. (A useful primer on architectural acoustics with a comprehensive, annotated bibliography.)

Acoustical Materials: Zwikker and Kosten, "Sound Absorbing Materials," Amsterdam, Elsevier, 1949. (Old, but one of the few books devoted exclusively to materials.)

Gonser, "Modern Materials," Volume 7, New York, Academic Press, 1970. (Contains a chapter on acoustical materials.) (Most of the previously listed references contain sections on acoustical materials, also.)

Berendt, "Airborne, Impact, and Structure Borne Noise Control in Multifamily Dwellings," Washington, D.C., U.S. Dept. of Housing and Urban Development, 1967. (A primer on the subject, and a good reference on constructions and materials for these purposes.)

Vibration Control: Harris and Crede, "Shock and Vibration Handbook," Second edition, New York, McGraw-Hill, 1976. (A comprehensive, modern reference.)

Crede, "Vibration and Shock Isolation," New York, Wiley, 1951. (One of the "classics" and still one of the best and most useful.)

HVAC Systems: *ASHRAE Guide and Data Book*, New York, ASHRAE, Biennial Publication. (The chapters on "Sound Control" are usually current and comprehensive.)

Sound Systems: Olson, "Acoustical Engineering," Brooklyn, Van Nostrand, 1957. (The best text available on electrical amplification systems, particularly as they apply to architectural acoustical design. Modern equipment developments have made it less useful for specific recommendations.)

Standards and Specifications: ASTM Standards and Publications.
ANSI Standards and Publications.
ASHRAE Standards and Publications.
ASME Standards and Publications.
SAE Standards and Publications.
Air Diffusion Council Standards and Publications.
Air Moving and Conditioning Association Publications and Standards.
International Standards Organization Standards and Publications.

Journals and Magazines: *Journal of the Acoustical Society of America*, New York. (A monthly, archival journal. Highly technical, but comprehensive and useful. Excellent indexes and cross-references.)

Sound and Vibration, Cleveland. (A free, monthly magazine, containing practical, useful articles on noise and vibration control.)

Noise Control Engineering, Poughkeepsie. (A bi-monthly publication of the Institute of Noise Control Engineering. Contains useful articles on all phases of noise control.)

Proceedings: *Noisexpo Proceedings*, Cleveland, Acoustical Publications, Inc. (Annual proceedings of conferences of noise control specialists in the USA.)

Noise-Con and Inter-Noise Proceedings, Poughkeepsie. (Annual proceedings of conferences sponsored by INCE, with papers covering a broad range of subjects.)

Additional Information: Major materials manufacturers provide, usually free of charge, helpful and usually reliable information.

The annual editions of *Sweet's Catalogs* contain a wide range of data on currently available materials, products, and constructions.

Each of the above references contains lists of additional references. The investigator interested in a comprehensive and exhaustive study of references will find considerable help in these books, particularly the indexes of the *Journal of the Acoustical Society of America.*

Various government agencies publish and make available at nominal cost, from time to time, information useful to designers and those interested in almost any phase of sound, noise, or vibration control.

Omission of any reference from the above list is not a reflection on its worth or usefulness; rather, we have included those likely to be most helpful to readers of this book. If time permits, the investigator may find it useful to look further, particularly to the references and bibliographies included in each of the above books and publications.

The International System of Units (SI)

Basic units (name, symbol, quantity)

meter m—length
kilogram kg—mass
second s—time
ampere A—electric current
kelvin K—thermodynamic temperature

Derived units

All other units are derived from basic and supplementary units. Some derived units have special names.

Viscosity Pa s (= N s/m^2 = kg/m s)

1 cP (centipoise) = 10^{-3} Pa s
1 lbf h/ft^2 = 0.172369 MPa s

Force newton N = kg m/s^2, N/m, pascal Pa = N/m^2

1 pdl = 0.138255 N
1 lbf = 4.44822 N
1 lbf/in^2 = 6894.76 Pa
1 kgf = kp = 9.80665 N
1 lbf/ft = 14.5939 N/m
1 dyne/cm = 1 (mN)/m (milli N Per m)
1 bar = 10^5 Pa
1 psi = 6.89476 kPa
1mm H$_2$O = 9.80665 Pa
1 in H$_2$O = 249.089 Pa
1 mm Hg = 133.322 Pa
1 at = kgf/cm^2 = 98.0665 kPa
1 atm = 101.325 kPa

Energy joule J = Nm = Ws
J/kg, J/kg°C (Note: °C = degree Celsius)

1 kWh = 3.6 MJ
1 Btu = 1.05506 kJ
1 Therm = 105.506 MJ
1 kcal = 4.1868 kJ
1 Btu/lb = 2.326 kJ/kg
1 Btu/lb°F = 4.1868 kJ/kg°C
1 ft pdl = 0.0421401 J
1 ft lbf = 1.35582 J

Power watt = J/s = N m/s
W/m², W/m²°C, W/m°C

1 Btu/h = 0.293071 W
1 kcal/h = 1.163 W
1 hp = 0.745700 kW
1 Ton refr. = 3.51685 kW
1 W/ft² = 10.7639 W/m²
1 Btu/h ft²°F = 5.67826 W/m²°C
1 Btu/h ft°F = 1.73073 W/m°C
1 Btu/h ft² (°F/in) = 0.144228 W/m°C

Length m, m/s

1 in = 25.4 mm
1 ft = 0.3048 m
1 yd = 0.9144 m
1 mile = 1.609344 km
1 ft/s = 0.3048 m/s
1 ft/min = 0.00508 m/s
1 mile/h = 0.44704 m/s
1 km/h = 0.277778 m/s

Area m²

1 in² = 0.00064516 m²
1 ft² = 0.09290304 m²
1 yd² = 0.836127 m²
1 acre = 4046.86 m²
       (= 0.404686 ha)
1 mile² = 2.58999 km²
1 a (are) = 10² m²
1 ha (hectare) = 10⁴ m²

Volume m³, m³/kg, m³/s
(Note: 1 liter = 10⁻³ m³)

1 ft³ = 28.3168 liters
1 US pint = 0.4732 liter

1 US gal = 3.7853 liters
1 ft^3/lb = 0.062428 m^3/kg
1 cfm = 0.471947 liter/s
1 cfm/ft^2 = 5.08000 liters/s m^2
                (air conditioning)

Mass kg, kg/m^3, kg/s, kg/s, kg/s m^2

1 oz = 28.3495 g
1 lb = 0.45359237 kg
1 long ton = 1.01605 Mg
              (= 1.01605 tonne)
1 lb/ft^3 = 16.0185 kg/m^3
1 g/cm^3 = 10^3 kg/m^3
1 lb/h = 0.00012599 kg/s
1 lb/h ft^2 = 0.0013562 kg/s m^2

Temperature

1 Rankine unit = 5/9 of Kelvin
(= 1 Fahrenheit = 5/9 of Celsius unit)

Rotational speed rev/s, rad/s
(Note: 1 rev = 2π rad)

in revolutions rev/s
1 rpm = 0.0166667 rev/s
in plane angle rad/s
1 rpm = 0.104720 rad/s
1 rev/s = 6.28318 rad/s

Index

Abrasion, 52
Absolute numbers, 6
Absolute silence, 14, 23
Absorbents, acoustical, 36, 37, 45, 48, 49,
    78, 79, 123, 209
  adhesive application of, 65
  amount and type of, 132
  applied, 131
  attaching and supporting of, 51
  location of, 78
  natural, 131
  porous, 46
Absorbers, space, 53
  diaphragmatic, 53
  selective, 138
Absorption, sound, 45, 48, 54, 131, 146
  mechanism of, 45
Absorption coefficients, sound, 46, 85, 86,
    133, 185, 203
Absorption curve, 51, 54
Absorptivity, total, 187, 203
Abuse, 52
Acceleration, formula, 161
  of gravity, 195
Accelerometer, 160, 205
Acoustic energy, 5, 32, 47, 48, 67, 77, 132
  definition, 1
Acoustic force, 5
Acoustical analysis, 84
Acoustical design, 18, 25, 67, 74, 84, 130,
    136, 141
  formulas, 57
Acoustical environment, 17, 23, 25, 41, 69,
    187, 197
Acoustical equation, 18, 76, 80

Acoustical materials, 5
  types of, 31
Acoustical treatment, 142, 206
  of mechanical equipment, 161
Acoustics, definition, 17
  units and dimensions of, 5
Addition, decibel, 8
Administrative controls, 188, 193, 224
Air, discharge, 182, 191
  intake, 182
  space, 50
Airborne sound, 27, 37, 41, 76, 96, 97, 113,
    120
  control of, 30, 32, 40, 211
  noise level, 160
Air conditioning systems, 76, 157, 170
Air-cooled condensers, 182
Aircraft noise, 188
  control of, 21, 97, 188, 198
Airflow, 47, 48, 50, 170
Airport noise, 188
Amplification, electronic, 30, 68
  components, 74
  ideal, 69
  natural, 67
  ratio, 148
  sound, 30, 67, 148
  vibration, 165
Amplifiers, 71, 151, 160
Amplitude, of motion, 5, 34
  definition, 4
  of vibration, 62
Analysis, acoustical, 84
  spectrum, 86
  system noise, 174

Anechoic, 28
  test chambers, 53, 138
Annoyance, 14, 19, 86
Architectural sound control, 17
A-scale, 19
Assembly line, 223
Athletic contest noise, 188
Attenuation, 27, 123, 126, 129, 152, 174,
        181, 187, 189, 192
  ceiling, 96
  of energy, 59
  of sound, 35, 56, 91, 170
Audible sound, 4, 12, 14
Audiometric tests, 200
Auditoriums, 27, 97, 136, 145
A-weighting, 38, 86
  network, 160

Background, noise, 67, 69, 75, 94
  criteria, 88
  sound level, 24, 130, 148, 194, 205
Baffles, 210, 217, 218, 219, 221
Balanced construction, 44, 123
Ballasts, 182
Barriers, partial-height, 44, 79, 120, 126,
        129, 161, 192, 224
  sound, 25, 30, 32, 41, 96, 123, 189, 211
Base, constructions, 105
  level, 195
Batts, 126, 225
Bel, 6
Bias, of the ear, 19
Binaural hearing, 14
Blasting, 151
Blocks, ceramic, 53
  concrete, as resonator, 55
  inertia, 166
  masonry, 55, 189
Blowers, lobed, 157
Board, gypsum, 129
Booths, foundry chipping, 222
Breaking up, of sound waves, 5, 48
Bridging, 51
Broadband, noise, 46
  sound, 24, 77, 130
B-scale, 19
Buffer spaces, 25, 26
Build-up, of energy, 80, 81, 186
  of noise, 28

C, middle, 11, 29
Cables, 181

Capacitance, 5
Carpets, 65, 131, 210
Case histories, 213, 214
Caulking, 43, 121
Ceilings, 27, 44, 53, 79, 96, 123, 126, 129,
        138, 177, 203, 209, 210
Centrifugal pump, 216
Chain saws, 222
Chipping booths, 222
Cleanliness, of absorbents, 49
Clouds, 138
Coefficients, 54
  noise reduction (NRC), 46, 132
  Sabine, 46
  sound absorption, 46, 85, 86, 132, 186,
        203
Coincidence, dip, 39
  effect, 35, 104
Combustion equipment, 77
Communication, of information, 13, 14
  of speech, 14, 67, 130
Components, electronic amplification, 74
Compression, 63, 65
Compressors, 182
  reciprocating refrigeration, 157
Concave surfaces, 29
Concert halls, 17, 30, 131, 138
Condensers, air-cooled, 181, 188
Configuration, 27, 137
Connections, flexible, 172
  resilient, 44, 78, 179
Conservation, hearing program, 200
Constrained layer, 65
Construction, balanced, 44, 123, 129
  base, 105
  diaphragm, 31
  double-wall, 36
  exterior, 25, 26, 95
  interior, 27, 96
  special high isolation, 120
Consumption of power, 191
Continuous open plenums, 123
Control, administrative, 188, 193
  impact, 30
  of mechanical equipment noise and
        vibration, 76, 153
  of noise, 18, 58
    along the propagation path, 189, 191
    at receiver, 189, 192
    at source, 188, 191
    environmental, 184, 187, 189

Control, of noise (*Cont.*)
  industrial, 190
  structure-borne, 30, 120
  of sound, 18, 67, 142, 211
  of vibration, 18, 58, 27, 30, 104
  structure-borne, 120
  panels, 75
  pulsation, 180
Controlled environment, 17
Conversion, of sound energy, 5, 46, 59, 68
Convex surfaces, 29
Conveyors, 191
  screw and bucket, 215
Cooling towers, 182, 188
Cost, 45, 49, 70, 96, 147, 157, 192, 200
Cost/benefit analysis, 213, 214
Coulomb damping, 61, 65, 67
Coupled spaces, 147, 177
Criteria, noise, 23, 85, 174
  environmental, 198
  industrial exposure, 195
  industrial vibration exposure, 197
Critical, building, 26
  damping, 66
    formula, 66
  listening, 27
  spaces, 26, 53, 120, 131, 146, 147, 164,
    170
Cross talk, 176
C-scale, 19
Cumulative exposure, 195
Curves, absorption, 51, 54
  noise criteria, 23, 87
Cycle, definition, 3

Dampers, 17
  pulsation, 157, 172
Damping, 37, 53, 58, 60, 104, 161, 167,
  192
  Coulomb, 61, 65, 67
  critical, 66
    formula, 66
  hysteresis, 61, 67
  materials, 65, 211
  ratio, related to loss factor η, 66
    formula, 66
  viscous, 61, 67
dBa, readings, 20
  values, 24
Dead, rooms, 28, 69
  spots, 29

Dead-load deflection, 165
Decay rate, 65
  related to loss factor η, 66
    formula, 66
Decibel (dB), 6, 19
  addition, 8
Deflection, dead-load, 165
  static, 164, 167
Density, 4, 48, 59
  formula, 4
Description, of environment, 18
Design, acoustical, 17, 25, 67, 74, 84, 130,
  136, 141
  formulas, 57
  of systems, 71
  procedure, 84
Diaphragm, constructions, 31
  floors, 44
Diaphragmatic absorbers, 53
Diffraction, 27, 29, 49, 71, 137, 138
  gratings, 29
Diffractors, 180
Diffusers, 172, 173
Diffusion, 27, 138
Dimension, 5, 27, 48, 54
Directional realism, 70
Directivity, 80, 184, 185
Discharge, air and steam, 181
Discrete sound, 28
Discrimination, 12
Dispersion pattern, 71, 149
Displacement, formula, 161
Dissipation, of energy, 59, 65, 79, 168
Distance, 91, 92
Distortion, 70, 151
Dividers, audiovisual, 146
Divisible spaces, 146
Doors, 43, 95, 129
  gaskets and perimeter seals, 123, 189
Double-wall construction, 36, 104
Doubling rate, 195
Drain pipes, 180
Draperies, 131
  movable, 131, 146
Driving, frequency, 164
  of energy, 67, 161
Ducts, 77, 175
  elbows in, 174
  liners, 175
  mufflers, 174, 175
  systems, 172

Ducts (*Cont.*)
  transfer, 129
  unlined, 175
Durability, 49
Dynamic loading, 64

Ear, muffs, 199
  plugs, 199
Echoes, 28, 30, 69, 73, 137, 146, 149, 209
Elastic, materials, 167, 160
  matter, 2
  media, 1, 2
Elasticity, 4, 35, 58
  modulus of, 4
    formula, 4
Elastomeric, edge gaskets, 129
  rollers, 192
Elastomers, 62, 63, 224
  bushings, 179, 180
  pads, 168
  tube liners, 225
Elbows, duct, 174, 175
Electric motors, 181
Electronic amplification, 30, 68
  components, 74
  equipment, 70
Elevators, 181
  hydraulic pumps, 157
Enclosure, 32, 44, 78, 161, 192
  partial, 159, 192
  reflective, 186
  sound in, 185, 190
    formula, 185
Energy, acoustic, 5, 32, 47, 67, 77, 132
  build-up, 80, 81
  conversion of, 5, 45, 60, 68
  dissipation of, 59, 60, 79, 168
  driving of, 67, 161
  kinetic, 5, 33, 55
    formula, 33
  potential, 5, 60
  storage of, 5
  thermal, 32
Entertainment noise, 188
Envelope of hearing, 10
Environment, acoustical, 17, 23, 25, 41, 69,
    197
  controlled, 17
  description of, 18
  evaluation of, 18

Environment (*Cont.*)
  ideal, 17
  measurement of, 18, 87
  physical, 17, 21, 22
Environmental noise, control, 181, 184, 187,
    189
  criteria, 198
Equal loudness contours, 11, 20, 21, 38
Equalization, 76, 152
Equation, acoustical, 18, 76
Equipment, electrical, 181, 187, 190, 204
  electronic amplification, 73
  imbalance, rotating, 160
  mechanical, 153, 187, 189, 190, 191
    acoustical treatment of, 161
    adjustment and maintenance, 158
    location of, 158
  noise and vibration, 76, 153, 160
    control of, 76, 153, 170
    incidental, 153
    inherent, 153
  ventilating, 76, 172, 177
Escalators, 181
  trusses, 181
Evaluation, of environment, 18
Exposure, 189, 193
  cumulative, 195
  industrial noise criteria, 195
  industrial vibration criteria, 195, 197
  long-term, 195
Exterior construction, 25, 26, 27
Exhaust mufflers, 191

Facings, 49
  material, 175
Fading, 151
Fall-off, 29
Fan, cooling, 157
  housings, 170, 175
  noise, 174
  propeller-type, 170
  systems, 170
  speed, 188
  tip speeds, 170
  velocity, 165
Feasible, definition, 213
Feedback, 70
Feeling, threshold of, 10, 14, 15
Felt, roofing, 179
FHA, 40, 103

Fiber, glass, 179, 192, 225
  quilted, 224
Fidelity, 69, 148
Field, houses, 188
  measurements, 204
  noise reduction, 201
  tests, 201
Fields, free, 80
  near, 184
  reverberant, 81
  sound, 187
Fillers, 65
Films, 50
Filters, sound system, 76, 152
Fire resistance, 27, 49
Flanking, 39, 41, 95, 121, 177
  definition, 42
  paths, 177
Flash tanks, 181
Flexible, connections, 172
  hangers, 178
Flexing, 52, 65
Floating floors, 120
Flocked surfaces, 5, 48
Floor/ceiling assemblies, 40, 103
Floors, 27, 31, 32, 39, 96, 102, 114, 147,
    164, 167, 177, 210
  diaphragm, 44
  floating, 120
  wood-joist, 40
Flow, laminar, 77, 172, 180
  maximum, 173
  noise, 77, 172, 179
  resistance, 48
  return air, 129
  traffic, 188
  turbulent, 77, 172, 179
  velocity, 77, 158, 172, 174, 179
Flutter, 28, 137, 209
Foundry chipping booths, 222
Flying, panels, 29
  reflectors, 138
Foams, 47, 52
  urethane, 52, 179, 192, 215, 216, 218,
    221
Focusing, 27, 67, 137, 209
Food processing noise, 220
Force, acoustic, 5, 33
  formula, 5, 33
Forcing frequency, 164

Free, fields, 80
  space, 80
Frequency, 29, 33, 34, 48, 71, 104, 161
  analysis, 19
    octave-band, 24, 205
  bands, 88, 201
    octave, 194, 201
  broadband, 77
  definition, 3
  driving, 164
  forcing, 62, 164
  fundamental, 153
  natural, 60, 164
    formula, 60, 164
  of resonator, 55
  ranges, 12, 29, 131
  resonant, 164
  response, 10, 152
  speech, 12, 148
Frictional losses, 47
Full-range reinforcement, 148
Fumes, 217
  effects of, 52
Furnaces, 191
Furnishings, 53, 131, 146, 147
Furring strips, 51
Fuzzy materials, 46, 131

Gaskets, 123
  elastomeric edge, 129
Geiger Thick-Plate Test, 65, 201
Glass, 95, 129, 225
  fiber, 192, 224, 225
Glass wools, 52, 53, 177, 179
Graphic level recorder, 203
Grilles, 172, 173
Gypsum board, 129

Halls, concert, 17, 30, 131, 138
  lecture, 30, 69, 148
Hangers, flexible, 178
Hard, 30
Hardness, of absorbents, 48, 49
Hearing, 9
  binaural, 14
  conservation of, 194
    program, 200
  damage, 195
  envelope of, 10
  fatigue, 14

Hearing (*Cont.*)
  loss of, 14, 94
  spaces, 25
  threshold of, 6, 10, 15
Honeycomb core panels, 36
Horns, 67, 71, 72
Hot spots, 29, 137
Housekeeping pads, 167
Housings, absorptive, 157
  fan, 175
  unlined, 175
HUD, 40
Human response, 6, 9, 24
  to vibration, 14
Humidity, effects of, 52
Hydraulic, elevator pumps, 157
  power unit, 217
Hysteresis, 4, 61, 67
  loop, 64
Hz, 3, 153

Ideal environment, 17
Impact, 27, 39, 52, 102, 114, 168, 181
  control of, 30
  insulation, 39
    Class (IIC), 40, 103
  isolation, 97
  noise rating (INR), 40, 103
impedance, 5, 40, 48, 67
  match, 37, 53, 67
  matching devices, 74, 151
  mismatch, 59
impulse, 3
Industrial noise, control, 184, 190
  exposure criteria, 195
  vibration exposure criteria, 197
Inertia blocks, 166
Information, communication of, 13, 24
Input, 71
  level control, 75, 151
Insulation, blanket, 126
  impact, 39
    Class (IIC), 40
  resilient, 179
  thermal, 192
Intensity, of sound, 5
  level, 7, 80
Interior construction, 96
Internal sound level, 26
Inverse square law, 4, 80

Isolation, 26, 32, 95, 97, 126, 129, 130,
    146, 161, 192, 199
  impact, 104
  shock, 79, 168
  sound, 33, 177
  vibration, 58, 78, 164, 210

Jerk, 79

Kinetic energy, 5, 33, 35
  formula, 33
"Kits," 157

"Lagging," pipe, 179, 192
Laws, inverse square, 4
  Limp Mass, 35
    formula, 35
  mass, 104
Lay-in panel, 44
Leaks, 34, 35, 39, 41, 94, 121, 123, 177,
    203, 209
Lecture halls, 30, 69, 148
Legal problems, 194
Levels, 85
  A-weighted, 205
  background sound, 24, 25, 88, 130, 148,
    194, 205
  base, 195
  intensity, 7
    formula, 7
  internal sound, 26
  linear, 205
  loudness, 10, 20
  masking, 130
  noise, 14, 79, 97, 170
    maximum, 188
  power, 7
  reference, 6
  signal, 14, 34, 69, 148
  sound power, 7, 184, 185, 186
    formula, 7, 184, 185, 186
  sound pressure, 7, 19, 20, 24, 34, 45, 52,
    69, 86, 95, 102, 130, 148, 175, 184,
    185, 186, 194, 202
    formula, 7, 184, 185, 186
Limp Mass Law, 35
  formula, 35
Limpness, 35, 50
Line, radiation, 184
  source, 80, 184

Linear network, 204, 205
Linings, absorbent, 79, 225
  duct, 173
Listening, critical, 27
  monaural, 152
Loading, 64
  dynamic, 64
Location, on site, 25
Longitudinal wave, 3
Long-term exposure, 195
Loss factor η, 66
  formula, 66
Loudness, 10, 19
  index, 21
  level, 10, 20
Louvers, 174, 189

Maintainability, of absorbents, 49
Masking, 13, 23, 130, 147, 189
Mass, 4, 33, 35, 54, 104, 166
  formula, 33, 104
  law, 104
Match, impedance, 37, 53, 67
Matching devices, impedance, 74, 151
Materials, acoustical, 5, 27, 31
  damping, 65, 211
  elastic, 167
  fuzzy, 46, 131
  lossy, 65
  porous absorptive, 46
  resilient, 52, 104
  resilient mounting, 62, 167
  sound-absorbing, 28, 37, 189
  surface, 27, 48, 49
  types of, 52
Matrix, 48, 65
Maximum flow, 173
Measurement, of environment, 18
  field, 204
  and tests, 201
Mechanical equipment, 153
  acoustical treatment of, 161
  adjustment and maintenance, 158
  electrical, 181
  location of, 158
  noise and vibration, 76, 153
    control of, 76, 78, 153, 170
    incidental, 153
    inherent, 153
  ventilating, 76, 172, 177

Medical problems, 190, 194
Medium, absorptive, 48
  elastic, 1, 2
  plastic, 17
Metal, panels, 223
  sheet, 224, 225
  wools, 52
Meter, sound level, 19, 38, 86, 87, 160, 202, 203
Microbar, 6
Microphones, 71, 74, 151, 205
  cable, 74, 151
  cardioid, 151
  directional, 151
  omnidirectional, 151
  outlets, 74, 152
Middle C, 11, 29
Mineral wools, 52, 177, 222, 225
Mismatch, impedance, 59
Mixing chambers, 147
Mock-ups, 214
Modes, 152
  room, 81, 152
Modulus, of elasticity, 4
  formula, 4
  shear, 37
Moisture resistance, 49
Monitoring, 75
Motors, 181
Mounting, 46, 51
  of equipment, 164
  of tile, 50, 51
  resilient, 62, 158, 164, 167, 170, 179, 181
  system requirements, 168
Movable partitions, 114, 123
Mufflers, 79, 157, 180
  duct, 174, 175
  exhaust, 191
  "splitter," 174
Muffs, ear, 199
"Mutes," 181

Natural, amplification, 67
  frequency, 60, 164
    formula, 60, 164
NC-number, 24, 87
Near field, 184
Networks, linear, 204, 205
  weighting, 19, 86, 160

Noise, 24, 103, 158, 194, 198
  background, 67, 69, 75, 94
  build-up of, 28
  control of, 18, 58
    aircraft, 21, 97, 198
    along the propagation path, 189, 191
    at the receiver, 189, 192
    at the source, 188
    environmental, 184, 187, 189
    industrial, 190
    sporting activities, 189
  Criteria, 23, 85, 88, 174, 175
    background, 88
    curve (NC-curve), 24, 87
  definition, 13
  effects on productivity, 197
  environmental, 181, 198
  extraneous, 96
  flow, 77, 172, 179
    radiation, 192
  food processing, 220
  generator, 98
  industrial, exposure criteria, 195
  internal, 95
  level, 14, 79, 97, 170
    perceived, 21
  mechanical equipment, control of, 76,
      153, 170
    incidental, 153
    inherent, 153
  out-shouting, 14, 69, 148
  "pink," 201
  pollution, 18
  radiation, 192
  Rating, Impact Noise (INR), 40, 103, 114
  Reduction, 28, 34, 57, 203
    field, 201
    formula, 34, 57, 135
    coefficient (NRC), 46, 132, 146
  regenerated, 175
  sources, 22, 25, 76, 94, 153, 170, 182,
      187, 190
  system analysis, 174
  thermal, 13
  traffic, 188
  "windage," 181
Noisiness, 19, 25
Noys, 21

Octave, 9, 130
  bands, 86, 194
  frequency increase, 34

Office areas, 24, 130
Oil, spray, 217, 219
Open-cell structures, 47
Open-plans, 146
  schools, 141
Open plenums, continuous, 123
Operating parameters, 157
Optimum reverberation times, 28, 135
Ordinances, zoning and noise, 198
Organ chambers, 147
Orientation, of spaces, 93
  of surfaces, 27, 28
  on site, 26, 91
Output, 71, 148
  level control, 75
Out-shouting, noise, 14, 69, 148
"Over-the-top" path, 123

P.A. systems, 68, 69, 190
Pads, housekeeping, 167
  resilient, 181
Pain, 9
  threshold of, 10
Paints, 42, 131
  textured, 5, 48
Panels, acoustical, 96
  flying, 29
  honeycomb core, 36
  lay-in, 44
  metal, 223, 225
  steel, 222
  vibrating, 80, 160, 205
    equations, 160, 161
Parameters, operating, 157
Partial-height barriers, 44, 79, 120, 161
Partitions, 26, 32, 95, 123, 129, 177
  movable, 114, 123
Path, 1, 18, 78, 211
  flanking, 177
  over-the-top, 123
  propagation, noise control along, 189, 191
  sound transmission, 18, 31, 44, 67, 77,
      178
Peak power, 74, 151
Peaks, 24
  absorptive, 55
Perceived noise level, 21
Personal protection, 193
Phon, 20
  scale, 20
Physical environment, 17, 21, 22
"Pink noise," 201

Pipe "lagging," 179
Piping systems, 178
Pitch, 11
  definition, 11
Plastic medium, 17
Plasticity, 4
Plenums, 170, 210
  continuous open, 123
  return air, 126
Plugs, ear, 199
PNdB, 21
Point source, 80, 184, 185
Porosity, 48
Porous absorptive materials, 46
Potential energy, 5, 60
Power consumption, 191
Power levels, sound, 7, 184, 185, 186, 201
  formula, 7, 184, 185
Press, stamping, 218, 219
Pressure pulse, 76, 79
Pressure, sound, 5, 34
  levels, 7, 19, 20, 24, 34, 45, 52, 69, 86,
    95, 102, 130, 148, 175, 184, 185,
    186, 202
  formulas, 7, 184, 185, 186
Privacy, 23, 130
Program, hearing conservation, 200
Propagation, of sound in elastic media, 2
Proportion, 5, 27, 137, 147
Protection measures, 198
  personal, 193
Pulsation control, 180
Pulverizer, Raymond Mill, 220
Pumps, centrifugal, 215
  hydraulic elevator, 157, 181
Pure tones, 12, 24

Quality, 12
  definition, 12
Quiet, 18, 25
  "models," 157

Racetracks, 189
Radiation, 31, 62, 78, 160, 161, 184, 211
  cylindrical, 80
  formulas, 160, 161
  from vibrating panels, 160
  hemispherical, 8, 184
  line, 184
  spherical, 80, 184

Ranges, frequency, 12, 29, 35, 131
  shooting, 188
Rate, doubling, 195
Ratio, signal-to-noise, 13, 74, 151
"Real time," 203
Receiver, 1, 10, 18, 45, 67, 78, 211
  noise control at, 189, 192
Receiving, end, 30, 145
  room, 201
Reciprocating refrigeration compressors,
    157
Recorder, graphic level, 203
Rectifiers, 182
Reduction, noise, 28, 34, 57, 203
  coefficient (NRC), 46, 132
  field, 201
  formula, 34, 57, 135
Reference intensity level, 7
Reference levels, 6
Reference pressure level, 7
Reflection, 45, 50, 58, 81, 131, 137, 138,
    189, 211
Reflective surfaces, 67, 79
Reflectors, 29, 68, 138
Refrigeration compressors, 157
Reinforcement, equipment, 70
  solid-state, 74
  full-range, 69, 148
  of sound, 27, 29, 67, 146
Resilient, attachment, 104, 178
  connections, 44, 78, 148, 179
  materials, 52, 104
    mounting, 62, 164
  mounting, 62, 158, 164, 167, 170, 181
  pads, 181
Resistance, 5, 48, 50
  flow, 48
Resonance, 165
  sharpness, 66
  formula, 66
Resonant absorbers, 55
Resonators, 55
  Helmholtz, 55
Response, frequency, 10, 152
  human, 6, 9, 24
  subjective, 9, 103
  to vibration, 14
Rest, 2
Reverberant fields, 81
Reverberation, 28, 45, 65, 69, 79, 131, 137,
    146, 149, 161, 209
  chamber, 54

Reverberation (*Cont.*)
  time, 57, 75, 135
    formula, 57, 136
    optimum, 128, 135
Reverberation decay tests, 203
Reverberation room method, 46
Roofs, 189, 225
"Room effect," 186
Room modes, 81
Rumble, 170, 181
Runners, 44, 51, 54, 129

Sabin, 46
Sabine, coefficients, 46
  formula, 57, 135, 203
  reverberation time formula, 46
Saws, chain, 222
Scale, 5
  macroscopic, 5
  molecular, 5
  sound pressure, 6
Schools, open-plan, 141, 147
Screens, 147, 189
Screw machines, automatic, 224
Sea of sound, 13, 28
Seats, 138, 147
Sending end, 30, 138, 147
Septum, 126, 129
Shape, 27, 137
Shear, 63, 65
  modulus, 37
  wave, 31, 35, 43
Shock, 64
  isolation, 79, 168
  transmission, 189
Shooting ranges, 188
Signal level, 14, 34, 69, 148
Signal-to-noise ratio, 13, 74, 151
Silence, absolute, 14, 23
Site, 25
  planning, 91
Size, 137
Skins, 37, 104
Slats, 29, 138
Sones, 21
Sonic booms, 197
Sound, absorption, 45, 48, 54, 131, 146
  absorption coefficient, 46, 85, 86, 133, 203
  airborne, 27, 30, 32, 37, 41, 76, 96, 97, 113, 120, 211

Sound (*Cont.*)
  amplification, 30, 67, 148
  audible, 4, 12, 14
  background, 24, 88, 130, 148, 194
  barriers, 25, 30, 32, 41, 96, 211
  control, 18, 67, 142, 211
    architectural, 17
  definition, 1
  discrete, 28
  energy density, 81
  fields, 79, 184, 187
  in enclosures, 79, 81, 184, 185, 190
    formula, 185
  in an environment, 17
  intensity of, 5
  isolation, 26, 33, 177
  level, internal, 26
  level meter, 19, 38, 86, 87, 160, 202, 203
  measurement of, 6
  motion of, 2
  nature of, 1
  objective, 10
  power level, 7, 80, 81
    formulas, 7, 81
  point, 80
  pressure, 5, 34
  pressure level, 7, 19, 20, 24, 34, 45, 52, 69, 81, 86, 95, 102, 130, 148, 175, 184, 185, 194, 201, 202, 203
    formulas, 7, 81, 82, 184, 185
  propagation of, in elastic media, 2
  reinforcement of, 27, 29, 67, 146
  sea of, 13, 28
  structure-borne, 27, 30, 42, 76, 102, 114, 120, 210
  systems, 68, 148
  Transmission Class (STC), 37, 38, 96, 130, 147, 187, 203
    contour, 103
    rating, 43, 85, 96, 114, 146
  transmission loss, 34, 96, 104, 187
    formula, 34
    measurement, 201
  transmission paths, 18, 31, 67, 77, 178
  unwanted, 13, 22, 24, 198
  velocity of, 4, 34, 59
    formula, 4, 34
  wanted, 13, 24, 67
  wave, 3
    breaking up of, 5, 48
Sound-absorbing materials, 28, 37

Soundproof, 18
Source, 1, 18, 45, 67, 77, 131, 137, 145,
    211
  line, 80, 184
  noise, 22, 25, 76, 94, 153, 170, 182, 190
  -path-receiver, 211
  point, 80, 184, 185
  random, 80
  room, 201
  weak, 29
Space absorbers, 53
Spaces, air, 50, 54
  buffer, 25, 26
  coupled, 147, 177
  critical, 26, 120, 131, 147, 164, 170
  design of specific, 141
  divisible, 146
  free, 80
  hearing, 25
  layout and arrangement of interior, 26, 93
  music, 55
  noisy, 25, 26, 70
  orientation of, 93
  quiet, 25, 26
  shape and configuration of, 27, 137, 147
  size of, 137
  speech, 146
  volume, 29
Speakers, 149
  controls, 151
  formula, 73, 149
  spacing, 72, 73, 149
Spectrum analysis, 86
Speech, communication, 14, 67, 130
  frequencies, 48, 69
  interference level, 130
  spaces, 146
Splices, 75, 151
"Splitter" muffler, 174
Spot-welding, 223
Springs, 62, 165
  collapsing, 64
  hardening, 64
  softening, 64
Squeaks, 29
Stadiums, 188
Stamping press, 218, 219
Standing waves, 82
Steel panels, 222
Stiffness, 4, 35, 54, 104
Strength, of absorbents, 49

Structure-borne sound, 27, 30, 42, 76, 102,
    114, 120, 192, 210
  control of, 30, 120
Structures, open-cell, 47
Studios, radio, 24, 55, 131, 177
  recording, 97, 120
Subjective response, 9, 103
Surface, area, of a cylinder, 80
  of a sphere, 80
Surface materials, 27, 48, 49
Surfaces, concave, 29, 209
  convex, 29
  orientation of, 27, 28
  reflective, 79, 67
  treating, 142
Systems, air-conditioning, 76, 157, 170
  design, 70, 71
  fan and duct, 170, 172
  filters in, 76
  gain, 68, 76, 148
  heating, 76, 170
  high-level (central speaker), 71, 149
  low-level (distributed speaker), 71, 149
  piping, 178
  plumbing, 77
  sound, 68, 145, 148
  split, 71
  variable volume, 173
  ventilating, 76, 172

Tapping test, 40, 103, 201
Temperature, effects of, 52
Templates and contours, 19
Terrain, 189
Tests, and measurements, 201
  audiometric, 200
  laboratory, 201
Textured paints, 5, 48
Theaters, 30, 54, 97, 148
Thermal, energy, 32
  noise, 13
Thickness, absorber, 48
Threshold, of feeling, 10, 14, 15
  of hearing, 6, 10, 15
  of pain, 10
Throttling devices, 78, 173
Tile, acoustical absorptive, 50, 96, 209
Tolerance, 15, 85, 190
Tones, pure, 12, 24
Towers, cooling, 181, 188

Tracks, race, 189
Traffic, flow, 188
 noise, 188
Transducers, 32, 45
Transformers, 74, 77, 151, 181
 dry-type, 157
Transmissibility, 164
Transmission, of sound, 27, 32, 192, 209
 energy, 5, 34, 58, 59
 paths, 18, 31, 44, 67, 77, 178
 shock, 189
 vibration, 182, 189
Transmission Class, Sound (STC), 37, 38,
 96, 187, 203
 rating, 43, 85, 96, 114
Transmission loss, ceiling, 126
 sound, 34, 96, 104, 187
 formula, 34
 measurement, 201
Transverse waves, 35
Treatment, acoustical, 142, 206
 of mechanical equipment, 161
Tuned, absorption, 138
 reflection, 138
 resonator, 55
 rooms, 75
Turbulent flow, 77, 172, 179, 180

Ultrasonics, 7
Units, of dimension, 5
Unwanted sound, 13, 22, 24, 198
 vibration, 160
Urethane foam, 179, 192, 215, 216, 218,
 221

Valves, 180
Vanes, absorbent turning, 174, 175
Variable volume systems, 173
Velocity, flow, 77, 158, 172, 179
 of sound, 4, 34, 59
 formula, 4, 34, 161
Ventilating equipment, 76, 172, 177
Vibrating, conveyors, 197
 panels, 80, 160
Vibration, 1, 210
 amplification of, 165
 control of, 18, 27, 30, 58, 104, 120
 criteria, 195, 197
 definition, 1

Vibration (*Cont.*)
 effects of, 15
 isolation, 58, 78, 164, 192, 210
 mechanical equipment, control of, 76,
 153, 170
 incidental, 153
  inherent, 153
 of resonator, 55, 56
 transmission, 81, 189
Vinyl, curtains, 217, 218, 219, 221, 224
 leaded, 179, 192, 215, 216
 loaded, 221
 transparent, 218, 224
Viscoelastic substances, 65, 104
Viscous damping, 61. 67
Voice-print, 12, 19
Voids, 47, 50, 123
Volume, of a space, 29

Walls, 27, 32, 42, 65, 71, 96, 104, 121,
 129, 145, 177, 189, 192, 209, 225
 operable, 123, 146
 performance, in place, 146
Wanted sound, 13, 24, 67
Wave effects, 167
 front, 4, 33, 53, 58, 137
 longitudinal, 3
 shear, 31, 35, 43
 standing, 81
 transverse, 35
Wavelength, 29, 48, 50, 55, 71
 definition, 3
Weather effects, 189
Wedges, 53
Weight, of absorbents, 49
 of construction, 104
 of equipment, 166, 167
Weighting networks, 19, 86, 160
Whole curve, 19
"Windage," 181
Windows, 43, 95, 129, 189, 210
Wood-joist floors, 40
Wools, glass, 52, 53, 177
 metal, 52
 mineral, 52, 177, 222, 225
Wythes, 104

Zero-level, of sound pressure, 6
Zoning, and noise ordinances, 198